HUMPHRY DAVY

HUMPHRY DAVY

POET AND PHILOSOPHER

T.E. THORPE

NONSUCH

First published 1896
Copyright © in this edition Nonsuch Publishing, 2007

Nonsuch Publishing
Cirencester Road, Chalford, Stroud, Gloucestershire, GL6 8PE
www.nonsuch-publishing.com

Nonsuch Publishing is an imprint of NPI Media Group

British Library Cataloguing in Publication Data.
A catalogue record for this book is available from the British Library.

ISBN 978 1 84588 382 9

Typesetting and origination by NPI Media Group
Printed in Great Britain

CONTENTS

INTRODUCTION TO THE
MODERN EDITION

IN THE EARLY NINETEENTH CENTURY Humphry Davy was arguably the most famous scientist in Europe. The son of an impoverished Cornish woodcarver, Davy rose to become a leading force in the reformed chemistry movement initiated by French chemist Antoine-Laurent de Lavoisier, and a pioneer in the new field of electrochemistry. Discoverer of sodium, potassium, barium, strontium, magnesium, calcium and boron, and of the anaesthetic properties of 'laughing gas' (nitrous oxide), he is also credited with inventing one of the first arc lamps, a miner's safety lamp and 'cathodic protection', which is still widely used to halt corrosion. He overturned Lavoisier's definition of acids as compounds of oxygen, and, within weeks of Italian physicist Alessandro Volta announcing the electric pile (an early type of battery) Davy produced incontrovertible evidence that it was a chemical reaction that produced electricity. This prompted him to explore electrolysis, which led him to isolate potassium. He also laid down some of the scientific foundations that enabled his protégé Michael Faraday to discover the laws of electrolysis. In addition to this Davy was an accomplished poet, known throughout his life to express his feelings in verse, and his impassioned and romantic spirit was unmistakable in his vision of science.

Davy was born on 17 December 1778 at Penzance in Cornwall. During his education at the grammar schools of Penzance and Truro he showed little sign of any interest in the sciences, or indeed any special talent in this area. His most noted ability appeared to be a certain skill in making verse translations from the classics and in writing stories. He left school just before the age of fifteen and the following year his father died, leaving his mother with substantial debts and five children to support. In order to clear her debts and provide for her family, Mrs Davy moved house, started a millinery business in Penzance and, in partnership with a young refugee Frenchwoman, took in lodgers to increase her income. The then sixteen-year-old Davy also awoke to his responsibilities as the eldest son, and through the help of a family friend became apprenticed to Dr John Borlase, a surgeon-apothecary in Penzance, where his work involved mixing potions in the laboratory. This opportunity was the start of a new life for Davy, who commenced upon a remarkably wide programme of self-education to prepare himself for an intended career in medicine, studying a number of subjects including theology, philosophy, literature, seven languages, and several sciences—including chemistry. Despite an affinity for literature and poetry, Davy's main passion proved to be for the study of electricity and chemistry, and he began to focus his studies more intensely within these areas. Around this time he made the acquaintance of two men of scientific accomplishment, Gregory Watt (the son of engineer James Watt), who was one of his mother's lodgers, and Davies Giddy, afterwards Gilbert, who was president of the Royal Society from 1827 to 1831. The latter recommended Davy to Dr Thomas Beddoes, who, in 1798, was establishing his Medical Pneumatic Institution in Bristol for the investigation of the medicinal properties of various gases, and he decided to install Davy as superintendent of the institution towards the end of that year.

This position enabled Davy to carry out experiments free from any restrictions and before long he had made his first significant discovery. His investigations into the properties of nitrous oxide led him to recommend its employment as an anaesthetic in minor surgical operations, and although this was largely ignored the breathing of the gas became fashionable within contemporary social gatherings, and his book, *Researches, Chemical and Philosophical, chiefly concerning Nitrous Oxide ... and its Respiration* (1799), secured his reputation as a chemist, as well as bringing the Pneumatic Institution to prominence. Davy's talents had now been noticed and he was asked in 1801 by the recently established Royal Institution in London to become an assistant lecturer in chemistry and director of the laboratory. After giving satisfactory proof of his abilities in a series of lectures upon galvanism he was appointed lecturer, later to be promoted to professor of chemistry in 1802. He spent the next few years moulding the institution into a centre for advanced research, particularly into electrochemistry, and as an arena for the delivery of popular demonstration lectures to audiences largely made up of the fashionable gentlemen and ladies of London society. It was during these years that he achieved many of his greatest accomplishments—the discovery of several alkali and alkaline-earth elements and contributions to the discoveries of the elemental nature of chlorine and iodine, for example, are some of the successes for which he is probably best remembered today.

By 1810 Davy's reputation was at its peak. As a lecturer he could command an audience of no less than 1,000 in the theatre of the Royal Institution, and his fame had spread much further than the confines of London. He gave a series of lectures in Dublin and was awarded an honorary degree of LL.D. by Trinity College, the only university distinction he ever received. He resigned from lecturing at the Royal Institution in 1812 in order to devote more time to original study, and with his wife, whom he had married the same year, and his assistant at the Royal Institution, Michael Faraday, he embarked upon a continental tour, during which he conducted research into various subjects and met with leading figures in science in France and Italy. Upon his return to England he was asked to take up the question of constructing a miner's safety lamp. He succeeded in inventing a lamp which provided light in the mines without risking the fires and explosions which had often been caused by flames or sparks coming into contact with methane gas, known as 'firedamp.' He achieved this at roughly the same time as engineer George Stephenson, who independently created a similar design, raising questions about the origins of the invention; however, Davy's lamp was praised by the mine owners who benefited from his design and rewarded his efforts with a gift of gold plate, as Davy did not wish to patent the design, feeling sufficiently recompensed by having contributed to the safety of the mining industry.

Davy received many more honours throughout his lifetime. He became a fellow of the Royal Society in 1803, serving as its president from 1820 to 1827; he was knighted in 1812 and created a baronet in 1818. It was after his resignation from the presidency of the Royal Society, due to ill health, that he again travelled in Europe, where he died in 1829. Largely remembered today for his impact within the world of scientific discovery, and most popularly for achievements such as the Davy Safety Lamp, *Humphry Davy: Poet & Philosopher* provides a fascinating account of not only this side of Davy but also an insight into his personal character—in particular his abiding love for literature, particularly poetry, something often overshadowed by his scientific persona. The poet Coleridge greatly admired Davy and declared 'had he not been the first chemist, he would have been the first poet of his age', while visiting his lectures to increase his own stock of metaphors. Davy's prose and poetry certainly betray a mind that was highly imaginative and creative, and the popularity of his lectures illustrates the gift he possessed for articulate exposition, despite his somewhat irritable temperament and peculiar manner. His determination to achieve fame, something that would at times lead to a tendency for petty jealousy, has been noted in contrast to his successor, Faraday, as have his apparent attempts to block his protégée's own progression. However, his interest in the 'cause of humanity' cannot be disputed, particularly in his invention of the miner's lamp, nor of course can the significant role he played in the social and intellectual world of London during this period.

PREFACE

FOR THE DETAILS OF SIR Humphry Davy's personal history, as set forth in this little book, I am mainly indebted to the well-known memoirs by Dr Paris and Dr John Davy. As biographies, these works are of very unequal value. To begin with, Dr Paris is not unfrequently inaccurate in his statements as to matters of fact, and disingenuous in his inferences as to matters of conduct and opinion. The very extravagance of his laudation suggests a doubt of his judgment or of his sincerity, and this is strengthened by the too evident relish with which he dwells upon the foibles and frailties of his subject. The insincerity is reflected in the literary style of the narrative, which is inflated and over-wrought. Sir Walter Scott, who knew Davy well and who admired his genius and his many social gifts, characterised the book as *"ungentlemanly"* in tone; and there is no doubt that it gave pain to many of Davy's friends who, like Scott, believed that justice had not been done to his character.

Dr Davy's book, on the other hand, whilst perhaps too partial at times—as might be expected from one who writes of a brother to whom he was under great obligations, and for whom, it is evident, he had the highest respect and affection—is written with candour, and a sobriety of tone and a directness and simplicity of statement far more effective than the stilted euphuistic periods of Dr Paris, even when he seeks to be most forcible. When, therefore, I have had to deal with conflicting or inconsistent statements in the two works on matters of fact, I have generally preferred to accept the version of Dr Davy, on the ground that he had access to sources of information not available to Dr Paris.

Davy played such a considerable part in the social and intellectual world of London during the first quarter of the century that, as might be expected, his name frequently occurs in the personal memoirs and biographical literature of his time; and a number of journals and diaries, such as those of Homer, Ticknor, Henry Crabb Robinson, Lockhart, Maria Edgeworth, and others that might be mentioned, make reference to him and his work, and indicate what his contemporaries thought of his character and achievements. Some of these references will be found in the following pages. It will surprise many Londoners to know that they owe the Zoological Gardens, in large measure, to a Professor of Chemistry in Albemarle Street, and that the magnificent establishment in the Cromwell Road, South Kensington, is the outcome of the representations, unsuccessful for a time, which he made to his brother trustees of the British Museum as to the place of natural history in the national collections. Davy had a leading share also in the foundation of the Athenaeum Club, and was one of its first trustees.

I am further under very special obligations to Dr Humphry D. Rolleston, the grand-nephew of Sir Humphry Davy, for much valuable material, procured through the kind co-operation of Miss Davy, the granddaughter of Dr John Davy. This consisted of letters from Priestley, Kirwan, Southey, Coleridge, Maria Edgeworth, Mrs Beddoes (Anna Edgeworth), Sir Joseph Banks, Gregory Watt, and others; and, what is of especial interest to his biographer, a large number of Davy's own letters to his wife. In addition were papers relating to the invention of the Safety Lamp. Some of the letters have already been published by Dr John Davy, but others now appear in print for the first time.

For the references to the early history of the Royal Institution I am mainly indebted to Dr Bence Jones's book. I have, moreover, to thank the Managers of the Institution for their kindness in giving me permission to see the minutes of the early meetings, and also for allowing me to consult the manuscripts and laboratory journals in their possession. These include the original records of Davy's work, and also the notes taken by Faraday of his lectures.

I have necessarily had to refer to the relations of Davy to Faraday, and I trust I have said enough on that subject. Indeed, in my opinion, more than enough has been said already. It is not necessary to belittle Davy in order to exalt Faraday; and writers who, like Dr Paris, unmindful of George Herbert's injunction, are prone to adopt an antithetical style in biographical narrative have, I am convinced, done Davy's memory much harm.

I regret that the space at my command has not allowed me to go into greater detail into the question of George Stephenson's relations to the invention of the safety lamp. I have had ample material placed at my disposal for a discussion of the question, and I am specially indebted to Mr John Pattinson and the Council of the Literary and Philosophical Society of Newcastle-upon-Tyne for their kindness in lending me a rare, if not unique, collection of pamphlets and reprints of newspaper articles which made their appearance when the idea of offering Davy some proof of the value which the coal owners entertained of his invention was first promulgated. George Stephenson's claims are not to be dismissed summarily as pretensions. Indeed, his behaviour throughout the whole of the controversy increases one's respect for him as a man of integrity and rectitude, conscious of what he thought due to himself, and showing only a proper assurance in his own vindication. I venture to think, however, that the conclusion to which I have arrived, and which, from the exigencies of space, is, I fear, somewhat baldly stated, as to the apportionment of the merit of this memorable invention, is just and can be well established. Stephenson *might* possibly have hit upon a safety lamp if he had been allowed to work out his own ideas independently and by the purely empirical methods he adopted, and it is conceivable that his lamp *might* have assumed its present form without the intervention of Davy; but it is difficult to imagine that an unlettered man, absolutely without knowledge of physical science, could have discovered the philosophical principle upon which the security of the lamp depends.

T.B.T.

May, 1896

I
PENZANCE: 1778–1798

Humphry Davy, the eldest son of "Carver" Robert Davy and his wife Grace Millett, was born on the 17th December, 1778.[1] His biographers are not wholly agreed as to the exact place of his birth. In the *Lives of Philosophers of the Time of George III* Lord Brougham states that the great chemist was born at Varfell, a homestead or " town-place" in the parish of Ludgvan, in the Mount's Bay, where, as the registers and tombstones of Ludgvan Church attest, the family had been settled for more than two hundred years.

Mr Tregellas, in his *Cornish Worthies* (vol. i., p. 247), also leaves the place uncertain, hesitating, apparently, to decide between Varfell and Penzance.

According to Dr John Davy, his brother Humphry was born in Market Jew Street, Penzance, in a house now pulled down, but which was not far from the statue of him that stands in front of the Market House of this town. Dr Davy further states that Humphry's parents removed to Varfell some years after his birth, when he himself was taken charge of by a Mr Tonkin.

The Davys originally belonged to Norfolk. The first member of the family that settled in Cornwall was believed to have acted as steward to the Duke of Bolton, who in the time of Elizabeth had a considerable property in the Mount's Bay. They were, as a class, respectable yeomen in fairly comfortable circumstances, who for generations back had received a lettered education. They took to themselves wives from the Eusticks, Adamses, Milletts, and other old Cornish families, and, if we may credit the testimony of the tombstones, had many virtues, were not over-given to smuggling or wrecking, and, for the most part, died in their own beds.

The grandfather of Humphry, Edmund Davy, was a builder of repute in the west of Cornwall, who married well and left his eldest son Robert, the father of the chemist, in possession of the small copyhold property of Varfell, to which reference has already been made. Robert, although a person of some capacity, seems to have been shiftless, thriftless, and lax in habits. In his youth he had been taught wood-carving, and specimens of his skill are still to be seen in and about Penzance. But he practised his art in an irregular fashion, his energies being mainly spent in field sports, in unsuccessful experiments in farming, and in hazardous, and for the most part fruitless, ventures in mining. At his death, which occurred when he was forty-eight, his affairs were found to be sadly embarrassed; his widow and five children were left in very straitened circumstances, and Varfell had to be given up.

Fortunately for the children, the mother possessed the qualities which the father lacked. Casting about for the means of bringing up and educating her family, she opened a milliner's shop in the town, in partnership with a French lady who had fled to England during the Revolution.

By prudence, good management, and the forbearance of creditors, she not only succeeded in rearing and educating her children, but gradually liquidated the whole of her husband's debts. Some years later, by an unexpected stroke of fortune, she was able to relinquish her business. She lived to a good old age, cheerful and serene, happy in the respect and affection of her children and in the esteem and regard of her townspeople. Such a woman could not fail to exercise a strong and lasting influence for good on her children. That it powerfully affected the character of her son Humphry, he would have been the first to admit. Nothing in him was more remarkable or more beautiful than his strong and abiding love for his mother. No matter how immersed he was in his own affairs, he could always find time amidst the whirl and excitement of his London life, amidst the worry and anxiety of official cares—or, when abroad, among the peaks of the None Alps or the ruins of Italian cities—to think of his far-away Cornish home and of her round whom it was centred. To the last he opened out his heart to her as he did to none other; she shared in all his aspirations, and lived with him through his triumphs; and by her death, just a year before his own, she was happily spared the knowledge of his physical decay and approaching end.

Davy was about sixteen years of age when his father died. At that time he was a bright, curly-haired, hazel-eyed lad, somewhat narrow-chested and undergrown, awkward in manner and gait, but keenly fond of outdoor sport, and more distinguished for a love of mischief than of learning.

Dr Cardew, of the Truro Grammar School, where, by the kindness of the Tonkins, he spent the year preceding his father's death, wrote of him that he did not at that time discover any extraordinary abilities, or, so far as could be observed, any propensity to those scientific pursuits which raised him to such eminence. "His best exercises were translations from the classics into English verse." He had previously spent nine years in the Penzance Grammar School under the tyranny of the Rev. Mr Coryton, a man of irregular habits and as deficient in good method as in scholarship. As Davy used to come up for the customary castigation, the worthy follower of Orbilius was wont to repeat:

Now, Master Davy,
Now, sir! I have 'ee
No one shall save 'ee—
Good Master Davy!

He had, too, an unpleasant habit of pulling the boys' ears, on the supposition, apparently, that their receptivity for oral instruction was thereby stimulated. It is recorded that on one occasion Davy appeared before him with a large plaster on each ear, explaining, with a very grave face, that he had "put the plasters on to prevent mortification." Whence it may be inferred that, in spite of all the caning and the ear-pulling, there was still much of the unregenerate Adam left in "good Master Davy."

Mr Coryton's method of inculcating knowledge and the love of learning, happily, had no permanent ill-effect on the boy. Years afterwards, when reflecting on his school-life, he wrote, in a letter to his mother:

After all, the way in which we are taught Latin and Greek does not much influence the important structure of our minds. I consider it fortunate that I was left much to myself when a child, and put upon no particular plan of study, and that I enjoyed much idleness at Mr Coryton's school. I perhaps owe to these circumstances the little talents that I have and their peculiar application.

If Davy's abilities were not perceived by his masters, they seemed to have been fully recognised by his school-fellows—to judge from the frequency with which they sought his aid in their Latin compositions, and from the fact that half the love-sick youths of Penzance employed him to write their valentines and letters. His lively imagination, strong dramatic power, and retentive memory combined to make him a good story and many an evening was spent by his comrades beneath the balcony of the Star Inn, in Market Jew Street, listening to his tales of wonder or horror, gathered from the "Arabian Nights" or from his grandmother Davy, a woman of fervid

mind stored with traditions and ancient legends, from whom he seems to
have derived much of his poetic instinct.

Those who would search in environment for the conditions which determine mental aptitudes will find it very difficult to ascertain what there was
in Davy's boyish life in Penzance to mould him into a natural philosopher.
At school he seems to have acquired nothing beyond a smattering of elementary mathematics and a certain facility in turning Latin into English verse.
Most of what he obtained in the way of general knowledge he picked up for
himself, from such books as he found in the library of his benefactor, Mr
John Tonkin. Dr John Davy has left us a sketch of the state of society in the
Mount's Bay during the latter part of the eighteenth century, which serves
to show how unfavourable was the soil for the stimulation and development
of intellectual power. Cornwall at that time had but little commerce; and
beyond the tidings carried by pedlars or ship-masters, or contained in the
Sherborne Mercury—the only newspaper which then circulated in the west of
England—it knew little or nothing of what was going on in the outer world.
Its roads were mostly mere bridle-paths, and a carriage was as little known in
Penzance as a camel. There was only one carpet in the town, the floors of the
rooms being, as a rule, sprinkled with sea-sand:

> All classes were very superstitious; even the belief in witches maintained its
> ground, and there was an almost unbounded credulity respecting the super-
> natural and monstrous … Amongst the middle and higher classes there was
> little taste for literature and still less for science, and their pursuits were rarely
> of a dignified or intellectual kind. Hunting, shooting, wrestling, cock-fighting,
> generally ending in drunkenness, were what they most delighted in. Smuggling
> was carried on to a great extent, and drunkenness and a low scale of morals
> were naturally associated with it.

Davy, an ardent, impulsive youth of strong social instincts, fond of excitement, and not over studious, seems, now that he was released from the
restraint of school-life, to have come under the influence of such surround-
ings. For nearly a year he was restless and unsettled, spending much of his
time like his father in rambling about the country and in fishing and shoot-
ing, and passing from desultory study to occasional dissipation. The death of
his father, however, made a profound impression on his mind, and suddenly
changed the whole course of his conduct. As the eldest son, and approaching
manhood, he seems at once to have realised what was due to his mother and
to himself. The circumstances of the family supplied the stimulus to exertion,

and he dried his mother's tears with the assurance that be would do all in his power for his brothers and sisters. A few weeks after the decease of his father he was apprenticed to Mr Bingham Borlase, an apothecary and surgeon practising in Penzance and at once marked out for himself a course of study and self-tuition almost unparalleled in the annals of biography, and to which he adhered with a strength of mind and tenacity of purpose altogether unlooked for in one of his years and of his gay and careless disposition. That it was sufficiently ambitious will be evident from the following transcript from the opening pages of his earliest notebook—a small quarto, with parchment covers, dated 1795:

1. Theology,
    or Religion,   } { taught by Nature;
    Ethics or Moral virtues } { by Revelation
2. Geography
3. My Profession
    1. Botany
    2. Pharmacy
    3. Nosology
    4. Anatomy
    5. Surgery
    6. Chemistry
4. Logic
5. Languages
    1. English
    2. French
    3. Latin
    4. Greek
    5. Italian
    6. Spanish
    7. Hebrew
6. Physics
    1. The doctrines and properties of natural bodies
    2. Of the operations of nature
    3. Of the doctrines of fluids
    4. Of the properties of organised matter
    5. Of the organisation of matter
    6. Simple astronomy
    7. Mechanics

8. Rhetoric and Oratory
9. History and Chronology
10. Mathematics

The notebook opens with "Hints Towards the Investigation of Truth in Religious and Political Opinions, composed as they occurred, to be placed in a more regular manner hereafter." Then follow essays "On the Immortality and Immateriality of the Soul"; "Body, Organised Matter"; on "Governments"; on "The Credulity of Mortals"; "An Essay to Prove that the Thinking Powers depend on the Organisation of the Body"; "A Defence of Materialism"; "An Essay on the Ultimate End of Being"; "On Happiness"; "On Moral Obligation."

These early essays display the workings of an original mind, intent, it may be, on problems beyond its immature powers, but striving in all sincerity to work out its own thoughts and to arrive at its own conclusions. Of course, the daring youth of sixteen who enters upon an inquiry into the most difficult problems of theology and metaphysics, with, what he is pleased to call, unprejudiced reason as his sole guide, quickly passes into a cold fit of materialism. His mind was too impressionable, however, to have reached the stage of settled convictions; and in the same notebook we subsequently find the heads of a train of argument in favour of a rational religious belief founded on immaterialism. Metaphysical inquiries seem, indeed, to have occupied the greater part of his time at this period; and his notebooks show that he made himself acquainted with the writings of Locke, Hartley, Bishop Berkeley, Hume, Helvetius, Condorcet, and Reid, and that he had some knowledge of the doctrines of Kant and the Transcendentalists.

That he thought for himself and was not unduly swayed by authority, is evident from the general tenour of his notes, and from the critical remarks and comments by which they are accompanied. Some of these are worth quoting:

Science or knowledge is the association of a number of ideas, with some idea or term capable of recalling them to the mind in a certain order.

By examining the phenomena of Nature, a certain similarity of effects is discovered. The business of science is to discover these effects, and to refer them to some common cause; that is to generalise ideas.

As his impulsive, ingenuous disposition led him, even to the last, to speak freely of what was uppermost in his mind at the moment, we may be sure that his elders, the Rev. Dr Tonkin, his good friend John Tonkin, and his grandmother Davy, with whom he was a great favourite, as he was with most old people, must have been considerably exercised at times with the metaphysical disquisitions to which they were treated; and we can well imagine that their patience was occasionally as greatly tried as that of the worthy member of the Society of Friends who wound up an argument with the remark, "I tell thee what, Humphry, thou art the most quibbling hand at a dispute I ever met with in my life." Whether it was in revenge for this sally that the young disputant composed the "Letter on the Pretended Inspiration of the Quakers" which is to be found in one of his early note-books, does not appear.

We easily trace in these early essays the evidences of that facility and charm of expression which a few years later astonished and delighted his audiences at the Royal Institution, and which remained the character-istic features of his literary style. These qualities were in no small degree strengthened by his frequent exercises in poetry. For Davy had early tasted of the Pierian spring, and, like Pope, may be said to have lisped in numbers. At five he was an *improvisatore*, reciting his rhymes at some Christmas gam-bols, attired in a fanciful dress prepared by a playful girl who was related to him. That he had the divine gift was acknowledged by no less an authority than Coleridge, who said that "if Davy had not been the first Chemist, he would have been the first Poet of his age." Southey also, who knew him well, said after his death, "Davy was a most extraordinary man; he would have excelled in any department of art or science to which he had directed the powers of his mind. He had all the elements of a poet; he only wanted the art. I have read some beautiful verses of his. When I went to Portugal, I left Davy to revise and publish my poem of 'Thalaba.'"

Throughout his life he was wont, when deeply moved, to express his feel-ings in verse; and at times even his prose was so suffused with the glow of poetry that to some it seemed altiloquent and inflated. Some of his first efforts appeared in the *Annual Anthology*, a work printed in Bristol in 1799, and edited by Southey and Tobin, and interesting to the book-hunter as one of the first of the literary "Annuals" which subsequently became so fashionable.

Davy had an intense love of Nature, and nothing stirred the poetic fire within him more than the sight of some sublime natural object such as a storm-beaten cliff, a mighty mountain, a resistless torrent, or so the spectacle which recalled the power and majesty of the sea. Not that he was insensible to the simpler charms of pastoral beauty, or incapable of sympathy with Nature in her softest, tenderest moods. But these things never seemed to move him as did some scene of grandeur, or some manifestation of stupendous natural energy.

The following lines, written on Fair Head during the summer of 1806, may serve as an example of how scenery when associated in his mind with the sentiments of dignity or strength affected him:

Majestic Cliff! Thou birth of unknown time,
Long had the billows beat thee, long the waves
Rush'd o'er thy hollow'd rocks, ere life adorn'd
Thy broken surface, ere the yellow moss
Had tinted thee, or the wild dews of heaven
Clothed thee with verdure, or the eagles made
Thy caves their aëry. So in after time
Long shalt thou rest unalter'd mid the wreck
Of all the mightiness of human works;
For not the lightning, nor the whirlwind's force,
Nor all the waves of ocean, shall prevail
Against thy giant strength, and thou shalt stand
Till the Almighty voice which bade thee rise
Shall bid thee fall.

In spite of a love-passage which seems to have provoked a succession of sonnets, his devotions to Calliope were by no means so unremitting as to prevent him from following the plan of study he had marked out for himself. His notebooks show that in the early part of 1796 he attacked the mathematics, and with such ardour that in little more than a year he had worked through a course of what he called "Mathematical Rudiments," in which he included "fractions, vulgar and decimal; extraction of roots; algebra (as far as quadratic equations); Euclid's elements of geometry; trigonometry; logarithms; sines and tangents; tables; application of algebra to geometry, etc."

In 1797 he began the study of natural philosophy, and towards the end of this year, when he was close on nineteen, he turned his attention to chemistry, merely, however, at the outset as a branch of his professional education, and with no other idea than to acquaint himself with its general principles. His good fortune led him to select Lavoisier's *Elements*—probably Kerr's translation, published in 1796—as his textbook. No choice could have been happier. The book is well suited to a mind like Davy's, and he could not fail to be impressed by the boldness and comprehensiveness of its theory, its admirable logic, and the clearness and precision of its statements.

From reading and speculation he soon passed to experiment. But at this time he had never seen a chemical operation performed, and had little or no acquaintance with even as much as the forms of chemical apparatus. Phials, wine-glasses, tea-cups, and tobacco-pipes, with an occasional earthen crucible, were all the paraphernalia he could command; the common mineral acids, the alkalis, and a few drugs from the surgery constituted his stock of chemicals. Of the nature of these early trials we know little. It is, however, almost certain that the experiments with seaweed, described in his two essays "On Heat, Light and the Combinations of Light" and "On the Generation of Phosoxygen and the Causes of the Colours of Organic Beings" (see p. 30), were made at this time, and it is highly probable that the experiments on land-plants, which are directly related to those on the *Fuci* and are described in connection with them, were made at the same period. That he pursued his experiments with characteristic ardour is borne out by the testimony of members of his family, particularly by that of his sister, who sometimes acted as his assistant, and whose dress too frequently suffered from the corrosive action of his chemicals. The good Mr Tonkin and his worthy brother, the Reverend Doctor, were also from time to time abruptly and unexpectedly made aware of his zeal. "This boy Humphry is incorrigible! He will blow us all into the air!" were occasional exclamations heard to follow the alarming noises which now and then proceeded from the laboratory. The well-known anecdote of the syringe which had formed part of a case of instruments of a shipwrecked French surgeon, and which Davy had ingeniously converted into an air-pump, although related by Dr Paris "with a minuteness and vivacity worthy of Defoe," is, in all probability, apocryphal. Nor has Lord Brougham's story, that his devotion to chemical experiments and "his dislike to the shop" resulted in a disagreement with his master, and that "he went to another in the same place," where "he continued in the same course," any surer foundation in fact.

Two or three circumstances conduced to develop Davy's taste for scientific pursuits, and to extend his opportunities for observation and experiment. One was his acquaintance with Mr Gregory Watt; another was his introduction to Mr Davies Gilbert (then Mr Davies Giddy), a Cornish gentleman of wealth and position, who lived to succeed him in the presidential chair of the Royal Society.

Gregory Watt, the son of James Watt, the engineer, by his second marriage, was a young man of singular promise who, had he lived, would—if we may judge from his paper in the *Philosophical Transactions*—almost certainly acquired a distinguished position in science. Of a weakly, consumptive habit, he was ordered to spend the winter of 1797 in Penzance, where he lodged with Mrs Davy, boarding with the family. Young Watt was about two years older than Davy, and had just left the University of Glasgow, "his mind enriched beyond his age with science and literature, with a spirit above the little vani-

ties and distinctions of the world, devoted to the acquisition of knowledge."
He remained in Penzance until the following spring, and by his example, and
by the generous friendship which he extended towards him, he developed and
strengthened Davy's resolve to devote himself to science. Davy's introduction
to Mr Gilbert, "a man older than himself, with considerable knowledge of
science generally, and with the advantages of a University education," was
also a most timely and propitious circumstance. According to Dr Paris:

> Mr Gilbert's attention was attracted to the future philosopher, as he was carelessly
> swinging over the hatch, or half-gate, of Mr Borlase's house, by the humorous
> contortions into which he threw his features. Davy it may be remarked, when
> a boy, possessed a countenance which even in its natural state was very far from
> comely; while his round shoulders, inharmonious voice and insignificant man-
> ner, were calculated to produce anything rather than a favourable impression:
> in riper years, he was what might be called 'good-looking,' although as a wit of
> the day observed, his aspect was certainly of the 'bucolic' character. The change
> which his person underwent, after his promotion to the Royal Institution, was
> so rapid that in the days of Herodotus, it would have been attributed to nothing
> less than the miraculous interposition of the Priestesses of Helen. A person, who
> happened to be walking with Mr Gilbert upon the occasion alluded to, observed
> that the extraordinary looking boy in question was young Davy, the carver's son,
> who, he added, was said to be fond of making chemical experiments.

Mr Gilbert was thus led to interest himself in the boy, whom he invited
to his house at Tredrea, offering him the use of his library, and such other
assistance in his studies as he could render. On one occasion he was taken
over to the Hayle Copper-House, and had the opportunity of seeing a well-
appointed laboratory:

> The tumultuous delight which Davy expressed on seeing, for the first time,
> a quantity of chemical apparatus, hitherto only known to him through the
> medium of engravings, is described by Mr Gilbert as surpassing all description.
> The air-pump more especially fixed his attention, and he worked its piston,
> exhausted the receiver, and opened its valves, with the simplicity and joy of a
> child engaged in the examination of a new and favourite toy.

It has already been stated that in the outset Davy attacked science as he
did metaphysics, approaching it from the purely theoretical side. As might be
surmised, his love of speculation quickly found exercise for itself, and within

four months of his introduction to the study of science he had conceived and elaborated a new hypothesis on the nature of heat and light, which he communicated to Dr Beddoes.

Dr Thomas Beddoes was by training a medical man, who in various ways had striven to make a name for himself in science. He is known to the chemical bibliographer as the translator of the Chemical Essays of Scheele, and at one time occupied the Chair of Chemistry at Oxford. The geological world at the end of the eighteenth century regarded him as a zealous and uncompromising Plutonist. His character was thus described by Davy, who in the last year of his life jotted down, in the form of brief notes, his reminiscences of some of the more remarkable men of his acquaintance:

> Beddoes was reserved in manner and almost dry; but his countenance was very agreeable. *He was cold* in conversation, and apparently much occupied with his own peculiar views and theories. Nothing could be a stronger contrast to his apparent coldness in discussion than his wild and active imagination, which was as poetical as Darwin's … On his deathbed he wrote me a most affecting letter, regretting his scientific aberrations.

One of Dr Beddoes's "scientific aberrations" was the inception and establishment of the Pneumatic Institution, which he founded with a view of studying the medicinal effects of the different gases, in the sanguine hope that powerful remedies might be found amongst them. The Institution, which was supported wholly by subscription, was to be provided with all the means likely to promote its objects—a hospital for patients, a laboratory for experimental research, and a theatre for lecturing.

In seeking for a person to take charge of the laboratory, Dr Beddoes bethought him of Davy, who had been recommended to him by Mr Gilbert. In a letter dated July 4th, 1798, Dr Beddoes thus writes to Mr Gilbert:

> I am glad that Mr Davy has impressed you as he has me. I have long wished to write to you about him, for I think I can open a more fruitful field of investigation than anybody else. Is it not also his most direct road to fortune? Should he not bring out a favourable result he may still exhibit talents for investigation, and entitle himself to public confidence more effectually than by any other mode. He must he maintained, but the fund will not furnish a salary from which a man can lay up anything. He must also devote his time for two or three years to the investigation. I wish you would converse with him upon the subject … I am sorry I cannot at this moment specify a yearly sum, nor can I say with certainty whether

all the subscribers will accede to my plan; most of them will, I doubt not. I have written to the principal ones, and will lose no time in sounding them all.

A fortnight later, Dr Beddoes again wrote to Mr Gilbert:

I have received a letter from Mr Davy since I wrote to you. He has oftener than once mentioned a *genteel maintenance*, as a preliminary to his being employed to superintend the Pneumatic Hospital. I fear the funds will not allow an ample salary; he must however be maintained. I can attach no idea to the epithet *genteel,* but perhaps all difficulties would vanish in conversation; at least I think your conversing with Mr Davy will be a more likely way of smoothing difficulties than our correspondence. It appears to me, that this appointment will bear to be considered as a part of Mr Davy's medical education, and that it will be a great saving of expense to him. It may also be the foundation of a lucrative reputation; and certainly nothing on my part shall be wanting to secure to him the credit he may deserve. He does not undertake to discover cures for this or that disease; he may acquire just applause by bringing out clear, though negative results. During my journeys into the country I have picked up a variety of important and curious facts from different practitioners. This has suggested to me the idea of collecting and publishing such facts as this part of the country will from time to time afford. If I could procure chemical experiments that bore any relation to organised nature, I would insert them. If Mr Davy does not dislike this method of publishing his experiments I would gladly place them at the head of my first volume, but I wish not that he should make any sacrifice of judgment or inclination.

Thanks to Mr Gilbert, the negotiation was brought to a successful issue. Mrs Davy yielded to her son's wishes, and Mr Borlase surrendered his indenture, on the back of which he wrote that he released him from "all engagements whatever on account of his excellent behaviour"; adding, "because being a youth of great promise, I would not obstruct his present pursuits, which are likely to promote his fortune and his fame." The only one of his friends who disapproved of the scheme was his old benefactor, Mr John Tonkin, who had hoped to have established Davy in his native town as a surgeon. Mr Tonkin was so irritated at the failure of his plans that he altered his will, and revoked the legacy of his house, which he had bequeathed to him.

---

1. In some biographical notices—e.g. in the *Gentlemam's Magazine*, xcix. pt. ii. 9—the year is given as 1779.

II

THE PNEUMATIC INSTITUTION, BRISTOL: 1798–1801

ON OCTOBER 2ND, 1798, DAVY set out for Clifton with such books and apparatus as he possessed, and the MSS of his essays on Heat and Light safely stowed away among his baggage. He was in the highest spirits, and full of confidence in the future. On his way through Okehampton he met the London coach decked with laurels and ribbons, and bringing the news of Nelson's victory of the Nile, which he interpreted as a happy omen. A few days after his arrival, he thus wrote to his mother:

> *October 11th*, 1798. Clifton.
>
> MY DEAR MOTHER,—I have now a little leisure time, and I am about to employ it in the pleasing occupation of communicating to you an account of all the *new* and *wonderful* events that have happened to me since my departure.
>
> I suppose you received my letter, written in a great hurry last Sunday, informing you of my safe arrival and kind reception. I must now give you a more particular account of Clifton, the place of my residence, and of my new friends Dr and Mrs Beddoes and their family.
>
> Clifton is situated on the top of a hill, commanding a view of Bristol and its neighbourhood, conveniently elevated above the dirt and noise of the city. Here are houses, rocks, woods, town and country in one small spot; and beneath us, the sweetly-flowing Avon, so celebrated by the poets. Indeed there can hardly be a more beautiful spot; it almost rivals Penzance and the beauties of Mount's Bay.
>
> Our house is capacious and handsome; my rooms are very large, nice and convenient; and, above all, I have an excellent laboratory. Now for the inhab-

itants, and, first, Dr Beddoes, who, between you and me, is one of the most original men I ever saw—uncommonly short and fat, with little elegance of manners, and nothing characteristic *externally* of genius or science; extremely silent, and in a few words, a very bad companion. His behaviour to me, however, has been particularly handsome. He has paid me the highest compliments on my discoveries, and has, in fact, become a convert to my theory, which I little expected. He has given up to me the whole of the business of the Pneumatic Hospital, and has sent to the editor of the *Monthly Magazine* a letter, to be published in November, in which I have the honour to be mentioned in the highest terms. Mrs Beddoes is the reverse of Dr Beddoes— extremely cheerful, gay and witty; she is one of the most pleasing women I have ever met with. With a cultivated understanding and an excellent heart, she combines an uncommon simplicity of manners. We are already very great friends. She has taken me to see all the fine scenery about Clifton; for the Doctor, from his occupations and his bulk, is unable to walk much. In the house are two sons and a daughter of Mr Lambton, very fine children, from five to thirteen years of age.

I have visited Mr Hare, one of the principal subscribers to the Pneumatic Hospital, who treated me with great politeness. I am now very much engaged in considering of the erection of the Pneumatic Hospital, and the mode of conducting it. I shall go down to Birmingham to see Mr Watt and Mr Keir in about a fortnight, where I shall probably remain a week or ten days; but before then you will again hear from me. We are just going to print at Cottle's; in Bristol, so that my time will be much taken up the ensuing fortnight in preparations for the press. The theatre for lecturing is not yet open; but, if I can get a large room in Bristol, and subscribers, I intend to give a course of chemical lectures, as Dr Beddoes seems much to wish it.

My journey up was uncommonly pleasant; I had the good fortune to travel all the way with acquaintances. I came into Exeter in a most joyful time, the celebration of Nelson's victory. The town was beautifully illuminated, and the inhabitants loyal and happy …

It will give you pleasure when I inform you that all my expectations are answered and that my situation is just what I could wish. But, for all this, I very often think of Penzance and my friends, with a wish to be there; however that time will come. We are some time before we become accustomed to new modes of living and new acquaintances.

Believe me, your affectionate son,
HUMPHRY DAVY.

Mrs Beddoes, of whom Davy speaks in such appreciative terms, was one of the many sisters of Maria Edgeworth. She seems to have possessed much of the intelligence, wit, vivacity, and sunny humour of the accomplished authoress of "Castle Rackrent"; and, by her charm of manner and her many social gifts, to have made her husband's house the centre of the literary and intellectual life of Clifton. Thanks to her influence, Davy had the good fortune to be brought into contact, at the very outset of his career, with Southey, Coleridge, the Tobins, Miss Edgeworth, and other notable literary men and women of his time, with many of whom he established firm and enduring friendships. He had always a sincere admiration for his fair patroness, and a grateful memory of her many acts of kindness to him at this period of his life. That she in turn had an esteem amounting to affection for the gifted youth is evident from the language of tender feeling and warm regard in which her letters to him are expressed. The sonnets accompanying these letters are couched in terms which admit of no doubt of the strength of her sentiments of sympathy and admiration, and some of the best efforts of his muse were addressed to her in return.

His work and prospects at the Pneumatic Institution are sufficiently indicated in the following letter to his friend and patron, Mr Davies Gilbert, written five weeks after his arrival at Clifton:

Clifton, *November* 12, 1798.

DEAR SIR,—I have purposely delayed writing until I could communicate to you some intelligence of importance concerning the Pneumatic Institution. The speedy execution of the plan will, I think, interest you both as a subscriber and a friend to science and mankind. The present subscription is, we suppose, nearly adequate to the purpose of investigating the medicinal powers of factitious airs; it still continues to increase, and we may hope for the ability of pursuing the investigation to its full extent. We are negotiating for a house in Dowrie Square, the proximity of which to Bristol, and its general situation and advantages, render it very suitable to the purpose. The funds will, I suppose, enable us to provide for eight or ten patients in the hospital, and for as many out of it as we can procure.

We shall try the gases in every possible way. They may be condensed by pressure and rarefied by heat. *Quere,*—Would not a powerful injecting syringe furnished with two valves, one opening into an air-holder and the other into the breathing chamber, answer the purpose of compression better than any other apparatus? Can you not, from your extensive stores of philosophy, furnish us with some hints on this subject? May not the non-respirable gases furnish a class of different stimuli? of which the oxy-muriatic acid gas [chlorine] would stand the highest, if we may judge from its effects on the lungs; then, probably, *gaseous oxyd of azote* [nitrous oxide?] and *hydro-carbonate* [the gases obtained by passing steam over red-hot charcoal].

I suppose you have not heard of the discovery of the native *sulphate of strontian* in England. I shall perhaps surprise you by stating that we have it in large quantities here. It had long been mistaken for sulphate of barytes, till our friend Clayfield, on endeavouring to procure the *muriate of barytes* from it by decomposition, detected the strontian. We opened a fine vein of it about a fortnight ago at the Old Passage near the mouth of the Severn.[1]

We are printing in Bristol the first volume of the 'West Country Collection,' which will, I suppose, be out in the beginning of January.

Mrs Beddoes … is as good, amiable, and elegant as when you saw her.

Believe me, dear Sir, with affection and respect, truly yours,
HUMPHRY DAVY.

The work alluded to in this letter made its appearance in the early part of 1799, under the title of "Contributions to Physical and Medical Knowledge, principally from the West of England; collected by Thomas Beddoes, MD." The first half of the volume, in accordance with the editor's promise, is occupied by two essays from Davy: the first "On Heat, Light, and the Combinations of Light, with a new Theory of Respiration"; the second "On the Generation of Phosoxygen (Oxygen Gas), and on the Causes of the Colours of Organic Beings."

To the student these essays have no other interest than is due to the fact that they are Davy's first contribution to the literature of science. No beginning could be more inauspicious. It is the first step that costs, and Davy's first step had well nigh cost him all that he lived for. As additions to knowledge they are worthless; indeed, a stern critic might with justice characterise them in much stronger language. Nowadays such writings would hopelessly damn the reputation of any young aspirant for scientific fame, for it is indeed difficult to believe, as we read paragraph after paragraph, that their author had any real conception of science, or that he was capable of understanding the need or appreciating the value of scientific evidence.

The essays are partly experimental, partly speculative, and the author apparently would have us believe that the speculations are entirely subservient to and dependent on the experiments. Precisely the opposite is the case. Davy's work had its origin in Lavoisier's *Traité Elémentaire*, almost the only textbook of chemistry he possessed. Lavoisier taught, in conformity with the doctrine of his time, that heat was a material substance, and that oxygen was essentially a compound body, composed of a simple substance associated with the matter of heat, or caloric. The young novitiate puts on his metaphysical shield and buckler; and with the same jaunty self-confidence that he assailed Locke and criticised Berkeley, enters the lists against this doctrine, determined, as he told Gregory Watt, "to demolish the French theory in half an hour."

After a few high-sounding but somewhat disconnected introductory sentences, and a complimentary allusion to "the theories of a celebrated medical philosopher, Dr Beddoes," he proceeds to put Lavoisier's question, "*La lumière, est-elle une modification du calorique, ou bien le calorique est-il une modification de la lumière?*" to the test of experiment. This he does by repeating Hawksbee's old experiment of snapping a gunlock "armed with an excellent flint" in an exhausted receiver. The experiment fails in his hands; such phenomena as he observes he misinterprets, and he at once concludes that light and heat have nothing essentially in common. "Nor can light be as some philosophers suppose, a vibration of the imaginary fluid ether. For even granting the existence of this fluid it must be present in the exhausted receiver as well as in atmospheric air; and if light is a vibration of this fluid, generated by collision between flint and steel in atmospheric air, it should likewise be produced in the exhausted receiver, where a greater quantity of ether is present, which is not the case." Since, then, it is neither an effect of caloric nor of an ethereal fluid, and "as the impulse of a material body on the organ of vision is essential to the generation of a sensation, *light is consequently matter of a peculiar kind*, capable when moving through space with the greatest velocity, of becoming the source of a numerous class of our sensations."

By experiments, faultless in principle but wholly imperfect in execution, he next seeks to show that caloric, or the matter of heat, has no existence. His reasoning is clear, and his conceptions have the merit of ingenuity, but any real acquaintance with the conditions under which the experiments were made would have convinced him that the results were untrustworthy and equivocal; and yet, in spite of the dubious character of his observations, he arrived at a theory of the essential nature of heat which is in accord with our present convictions, and which he states in the following terms:

Heat, or that power which prevents the actual contact of the corpuscles of bodies, and which is the cause of our peculiar sensations of heat and cold, may be defined a peculiar motion, probably a vibration, of the corpuscles of bodies, tending to separate them.

This conception of the nature of heat did not, of course, originate with him, and it was a question with his contemporaries how far he was influenced by Rumford's work and teaching. On this point Dr Beddoes's testimony is direct and emphatic. He says:

> The author [Davy] derived no assistance whatever from the Count's ingenious labours. My first knowledge of him arose from a letter written in April 1798, containing an account of his researches on heat and light; and his first knowledge of Count Rumford's paper was conveyed by my answer. The two Essays contain proofs enough of an original mind to make it credible that the simple and decisive experiments on heat were independently conceived. Nor is it necessary, in excuse or in praise of his system, to add, that, at the time it was formed, the author was under twenty years of age, pupil to a surgeon-apothecary, in the most remote town of Cornwall, with little access to philosophical books, and none at all to philosophical men.

Having thus, with Beddoes, expunged caloric from his chemical system, Davy proceeds to elevate the matter of light into its place. According to Lavoisier oxygen gas was a compound of a simple substance and caloric; Davy seeks to show that it is a compound of a simple substance and light. He objects to the use of the word "gas," since, according to French doctrine, it is to be taken as implying not merely a state of aggregation but a combination of caloric with another substance, and suggests therefore that what was called oxygen gas should henceforth be known as *phosoxygen*. His "proofs" that oxygen is really a compound of a simple substance with "matter in a peculiar state of existence" are perhaps the most futile that could be imagined. Charcoal, phosphorus, sulphur, hydrogen, and zinc were caused to burn in oxygen; *light* was evolved, oxides were formed, *and a deficiency of weight was in each case observed*. He regrets, however, that he "possessed no balance sufficiently accurate to determine exactly the deficiency of weight from the light liberated in different combustive processes."

> From these experiments, it appears that in the chemical process of the formation of many oxyds and acids, light is liberated, the phosoxygen and combustible base consumed, and a new body formed ... Since light is liberated in these processes, it is evident that it must be liberated either from the phosoxygen or from the combustible body ... If the light liberated in combustion be supposed (according to Macquer's and Hutton's theories) to arise from the combustible body, then phosoxygen must be considered as a simple

substance; and it follows on this supposition, that whenever phosoxygen combines with combustible bodies, either directly or by attraction from any of its combinations, light must be liberated, which is not the case, as carbon iron and many other substances, may be oxydated by the decomposition of water without the liberation of light.

Davy is here on the horns of a dilemma, but he ignores the difficulty, and, with characteristic "flexibility of adaptation," proceeds to offer synthetical proofs "that the presence of light is absolutely essential to the production of phosoxygen." The character of the "proofs" is sufficiently indicated by the following extracts:

When pure oxyd of lead is heated as much as possible, included from light, it remains unaltered; but when exposed to the light of a burning-glass, or even of a candle, phosoxygen is generated and the metal revivified.

Oxygenated muriatic acid [chlorine] is a compound of muriatic acid, oxygen and light, as will be hereafter proved. The combined light is not sufficient to attract the oxygen from the base [muriatic acid] to form phosoxygen; but its attraction for oxygen renders the [oxygenated muriatic] acid decomposable. If this acid be heated in a close vessel and light excluded no phosoxygen is formed; but if it be exposed to the solar light, phosoxygen is formed; the acid loses its oxygen and light and becomes muriatic acid.

A plant of Arenaria Tenuifolia planted in a pot filled with very dry earth, was inserted in carbonic acid, under mercury. The apparatus was exposed to the solar light, for four days successively, in the month of July. By this time the mercury had ascended considerably. The gas in the vessel was now measured. There was a deficiency of one-sixth of the whole quantity. After the carbonic acid was taken up by potash, the remaining quantity, equal to one-seventh of the whole, was *phosoxygen almost pure*. From this experiment, it is evident that carbonic acid is decomposed by two attractions; that of the vegetable for carbon and of light for oxygen: the carbon combines with the plant, and the light and oxygen combined are liberated in the form of phosoxygen

The accounts which Davy gives of his experiments, as well as of the phenomena which he professes to have observed, may awaken an uneasy doubt as to his absolute integrity; for, it is hardly necessary to point out, he could not possibly have obtained the results which he describes. The presence or absence of light in no wise affects the decomposition by heat of minium; chlorine, as he himself subsequently established contains no oxygen; and a plant is incapable of decomposing pure undiluted carbonic acid, even in the brightest sunshine. But the work of a youth of nineteen, imaginative sanguine, and impetuous with no training as an experimentalist, and with only a limited access to scientific memoirs cannot be judged by too severe a canon. The faculty of self-deception, even in the largest and most receptive minds, often in those of matured power and ripened experience, is boundless. Davy himself affords an exemplification of the truth of his own words, written years afterwards: "The human mind is always governed not by what it knows, but by what it believes; not by what it is capable of attaining, but by what it desires."

It is not necessary to show how the presumptuous youth drove his hobby with all the reckless daring of a Phaeton. Phlogiston and oxygen had in turn been the central conceptions of theories of chemistry; phosoxygen was to supplant them. It was to explain everything—the blue colour of the sky, the electric fluid, the Aurora Borealis, the phenomena of fiery meteors, the green of the leaf; the red of the rose, and the sable hue of the Ethiopian; perception, thought, and happiness; and why women are fairer than men. But Jupiter, in the shape of a Reviewer, soon hurled the adventurous boy from the giddy heights to which he had soared. The "West Country Collection" received scant sympathy from the critics, and the phosoxygen theory was either mercilessly ridiculed, or treated with contempt.

There is no doubt that Davy keenly felt the position in which he now stood. His pride was humbled, and the humiliation was as gall and wormwood. The vision of fame which his ardour had conjured up on the top of the Bristol coach—was it all a baseless fabric, and its train of honours and emoluments an insubstantial pageant? All he could plead was that his critics had not understood that these experiments were made when he had studied chemistry only four months, when he had never seen a single experiment executed, and when all his information was derived from Nicholson's *Chemistry* and Lavoisier's *Elements*. But his good sense quickly came to his rescue. After the first feelings of anger and mortification had passed, he recognised the justice of his punishment, much as he might resent the mode in which it was inflicted. How keen was the smart will appear from the following reflection, written in the August of the year in which the essays were published:

When I consider the variety of theories that may be formed on the slender foundation of one or two facts, I am convinced that it is the business of the true philosopher to avoid them altogether. It is more laborious to accumulate facts

than to reason concerning them; but one good experiment is of more value
than the ingenuity of a brain like Newton's.

About the same time he wrote:

> I was perhaps wrong in publishing, with such haste, a new theory of chemis-
> try. My mind was ardent and enthusiastic. I believed that I had discovered the
> truth. Since that time my knowledge of facts is increased —since that time I
> have become more sceptical.

In the October of the same year he wrote:

> Convinced as I am that chemical science is in its infancy, that an infinite variety
> of new facts must be accumulated before our powers of reasoning will be suf-
> ficiently extensive, I renounce my own particular theory as being a complete
> arrangement of facts it appears to me now only as the most apt.

By the end of the year the repentance was complete, and recantations followed. In a letter which appeared in Nicholson's Journal in February, 1800,
he corrects some of the errors into which he had fallen, and says, "I beg to
be considered as a sceptic with regard to my own particular theory of the
combinations of light, and theories of light in general." To the end of his
days Davy never forgot the lesson which his earliest effort had taught him;
and there is no question that the memory of it acted as a salutary check on
the exuberance of his fancy and the flight of his imagination. The wound to
his self was, however, never wholly healed. Nothing annoyed him more than
any reference to Beddoes's book, and he declared to Dr Hope that he would
joyfully relinquish any little glory or reputation he might have acquired by
his later researches were it possible to withdraw his share in that work and to
remove the impression he feared it was likely to produce.

And yet, in spite of the unqualified censure with which they were received,
and of the severe condemnation of them by their own author, we are disposed
to agree with Dr Davy that posterity will not suffer these essays to he wholly
blotted out from the records of science. That the experimental part was for
the most part radically bad, that the generalisation was hasty and presumptuous, and the reasoning imperfect, cannot be gainsaid. But, withal, the essays
display some of Davy's best and happiest characteristics. There is dignity of
treatment and a sense of the nobility of the theme on which he is engaged; the
literary quality is admirable; there is clearness of perception and perspicuity of

statement; the facts as he knew them—or as he thought he knew them—are marshalled with ingenuity and with a logical precision worthy of his model and teacher Lavoisier; his style is sonorous and copious, even to redundancy—some of the periods indeed glow with all the fervour and richness of his Royal Institution lectures. However wild and visionary his speculations may seem, minds like those of Coleridge and Southey were not insensible to the intrinsic beauty of some of his ideas. His theory of respiration might not be true, but it had at least the merit of poetic charm in its consequence that the power and perspicacity of a thinker had some relation to the amount of light secreted by his brain. Even good old Dr Priestley, whose Pegasus could never be stirred beyond the gentlest of ambles, tells us in the Appendix to his *Doctrine of Phlogiston Established* that Mr H. Davy's essays had impressed him with a high opinion of the philosophical acumen of their author. "His ideas were to me new and very striking; but," he adds, with a caution that was not habitual, "they are of too great consequence to be decided upon hastily."

Among the letters entrusted to me is Priestley, which must have been particularly to the young man. It is as follows:

> Northumberland, *Oct.* 31, 1801.
>
> Sir,—I have read with admiration your excellent publications, and have received much instruction from them. It gives me peculiar satisfaction that, as I am far advanced in life, and cannot expect to do much more, I shall leave so able a fellow-labourer of my own country in the great fields of experimental philosophy. As old an experimenter as I am, I was near forty before I made any experiments on the subject of Air, and then without, in a manner, any previous knowledge of chemistry. This I picked up as I could, and as I found occasion for it, from books. I was also without apparatus, and laboured under many other disadvantages. But my unexpected success induced the friends of science to assist me, and then I wanted for nothing. I rejoice that you are so young a man and perceiving the ardour with which you begin your career, I have no doubt of your success.
>
> My son, for whom you express a friendship, and which he warmly returns, encourages me to think that it may not be disagreeable to you to give me information occasionally of what is passing in the philosophical world, now that I am at so great a distance from it, and interested, as you may suppose, in what passes in it. Indeed, I shall take it as a great favour. But you must not expect anything in return. I am here perfectly insulated, and this country furnishes but few fellow-labourers, and these are so scattered, that we can have but little communication with each other, and they are equally in want of

information with myself. Unfortunately too, correspondence with England is very slow and uncertain, and with France we have not as yet any intercourse at all, though we hope to have it soon …

I thank you for the favourable mention you so frequently make of my experiments, and have only to remark that in Mr Nicholson's Journal you say that the conducting power of charcoal was first observed by those who made experiments on the pile of Volta; whereas it was one of the earliest that I made, and gave an account of in my History of Electricity, and in the Philosophical Transactions. And in your treatise on the Nitrous Oxide p. 55 you say, and justly, that I concluded this air to be lighter than that of the atmosphere. This, however, was an error in the printing that I cannot account for. It should have been *alkaline air*, as you will see the experiment necessarily requires.

> With the greatest esteem, I am Sir, yours sincerely
> J. PRIESTLEY.

In Davy's next contribution, "On the Silex composing the Epidermis, or External Bark, and contained in other parts of certain Vegetables," published in Nicholson's Journal in the early part of 1800, we find the evidence of a chastened and contrite spirit. The theme is humble enough, and the language as sober and sedate as that of Mr Cavendish. The chance observation of a child that two bonnet-canes rubbed together in the dark produced a luminous appearance, led him to investigate the cause, which he found to reside in the crystallised silica present in the epidermis. Reeds and grasses, and the straw of cereals, were also found to be rich in silica, from which he concludes that "the flint entering into the composition of these hollow vegetables may be considered as analogous to the bones of animals; it gives to them stability and form, and by being situated in the epidermis more effectively preserves their vessels from external injury." It is doubtful, however, whether the rigidity of the stems of cereals is wholly due to the silica they contain.

From a letter to Mr Davies Gilbert, dated April 10th, 1799, we learn that he had now begun to investigate the effects of gases in respiration. In the early part of the year he had removed to a house in Dowry Square, Clifton, where he had fitted up a laboratory. After thanking his friend for his critical remarks on his recently published essays, he says:

> Your excellent and truly philosophic observations will induce me to pay greater attention to all my positions … I made a discovery yesterday which proves how necessary it is to repeat experiments. The gaseous oxide of azote is perfectly

respirable when pure. It is never deleterious but when it contains nitrous gas. I have found a mode of obtaining it pure, and I breathed today, in the presence of Dr Beddoes and some others, sixteen quarts of it for near seven minutes. It appears to support life longer than even oxygen gas, and absolutely intoxicated me. Pure Oxygen gas produced no alteration in my pulse, nor any other material effect; whereas this gas raised my pulse upwards of twenty strokes, made me dance about the laboratory as a madman, and has kept my spirits in a glow ever since, is not this a proof of the truth of my theory of respiration? I this gas contains more light in proportion to its oxygen than any other, and I hope will prove a most valuable medicine.

We have upwards of eighty out-patients in the Pneumatic Institution, and are going on wonderfully well.

This observation of the respirability of nitrous oxide, and of the effects of its inhalation, was quickly confirmed. Southey, Coleridge, Tobin (the dramatist), Joseph Priestley, the son of the chemist, the two Wedgwoods, and a dozen other people of lesser note were induced to breathe the gas and to record their sensations, The discovery was soon noised abroad; Dr Beddoes dispatched a short note to Nicholson's Journal, and the fame of the Pneumatic Institution went up by leaps and bounds.

Maria Edgeworth, who was at the time on a visit to her sister, thus writes:

A young man, a Mr Davy, at Dr Beddoes', who has applied himself much to chemistry, has made some discoveries of importance, and enthusiastically expects wonders will be performed by the use of certain gases, which inebriate in the most delightful manner, having the oblivious effects of Lethe, and at the same time giving the rapturous sensations of the Nectar of the Gods! Pleasure even to madness is the consequence of this draught. But faith, great faith, is I believe necessary to produce any effect upon the drinkers, and I have seen some of the adventurous philosophers who sought in vain for satisfaction in the bag of *Gaseous Oxyd,* and found nothing but a sick stomach and a giddy head.

Laughing-gas, indeed threatened to become, like Priestley's dephlogisticated air, "a fashionable article in luxury." Monsieur Fiévée, in his *Lettres sur l'Angleterre,* 1802, names it in the catalogue of follies to which the English were addicted, and says the practice of breathing it amounted to a national vice!

Davy had no sooner discovered that the gas might be respired, than he proceeded to attack the whole subject of the chemistry of the oxides of nitrogen, and of nitrous oxide in particular, and after ten months of incessant labour he put together the results of his observations in an octavo volume, entitled, "Researches, Chemical and Philosophical, chiefly concerning Nitrous Oxide, or Dephlogisticated Nitrous Air, and its Respiration. By Humphry Davy, Superintendent of the Medical Institution." The book appeared in the summer of 1800, and immediately re-established its author's character as an experimentalist. Thomson, in his *History of Chemistry*, says of it: "This work gave him at once a high reputation as a chemist, and was really a wonderful performance, when the circumstances under which it was produced are taken into consideration." In spite, however, of the eulogies with which it was welcomed, its sale was never very extensive, and a second edition was not required. In fact, the work as a whole was hardly calculated to attract the general public, whose only concern with laughing-gas was in its powers as an exhilarant. Indeed, this aspect of the question is not wholly lost on Davy himself, who is careful to point out that "if the pleasurable effects or medical properties of the nitrous oxide should ever make it an article of general request, it may be procured with much less time, labour, and expense than most of the luxuries, or even necessaries, of life" and in a footnote he adds, "A pound of nitrate of ammonia costs 5s 10d (its present price is 9d!). This pound, properly decomposed, produces rather more than 34 moderate doses of the air, so that the expense of a dose is about 2d. What fluid stimulus can be procured at so cheap a rate?"

To the chemical student the book had, and still has, many features of interest. It contains a number of important facts, based on original and fairly accurate observation In the arrangement of these facts "I have been guided as much as possible," says their author "by obvious and simple analogies only. Hence, I have seldom entered into theoretical discussions, particularly concerning light, heat, and other agents, which are known only by isolated effects. Early experience has taught me the folly of hasty generalisation." The work is divided into four main sections. The first chiefly relates to the production of nitrous oxide, and the analysis of nitrous gas and nitrous acid. He minutely studies the mode of decomposition of ammonium nitrate (first observed by Berthollet), and shows that it is an endothermic phenomenon varying in character with the temperature and manner of heating. He is thus led to offer the following *Speculations on the Decompositions of Nitrate of Ammonia*:

All the phenomena of chemistry concur in proving that the affinity of one body, A, for another, B, is not destroyed by its combination with a third, C, but only modified; either by condensation or expansion, or by the attraction of C for B. On this principle the attraction of compound bodies for each other must be resolved into the reciprocal attraction of their constituents and consequently the changes produced in them by variations of temperature explained from the alterations produced in the attractions of those constituents.

The singular property possessed by ammonium nitrate of decomposing in several distinct modes according to the rapidity of heating and the temperature to which the substance is raised, first indicated by Davy, has been minutely studied by M. Berthelot, who has shown that this comparatively simple salt may be decomposed in as many as six different ways. It may be (1) dissociated into gaseous nitric acid and ammonia; (2) decomposed into nitrous oxide and water; (3) resolved into nitrogen, oxygen, and water, (4) or into nitric oxide, nitrogen, and water, (5) or into nitrogen, nitrogen peroxide, and water; or lastly (6) under the influence of spongy platinum, it may be resolved into gaseous nitric acid, nitrogen, and aqueous vapour. These different modes of decomposition may be distinct or simultaneous; or, more exactly, the predominance of any one of them depends on relative rapidity and on the temperature at which decomposition is produced. This temperature is not fixed, but is itself subordinate to the rapidity of heating (cf. Berthelot's *Explosives and Their Power*, translated by Hake and Macnab). The assertion of De la Metherie, that the gas produced by the solution of platinum in nitro-muriatic acid was identical with the dephlogisticated nitrous air of Priestley (nitrous oxide), led Davy to examine the gaseous products of this reaction more particularly. He had no difficulty in disproving the statement of the French chemist; but his observations, although accurate, led him to no definite conclusion. "It remains doubtful," he says, "whether the gas consists simply of highly oxigenated muriatic acid and nitrogen, produced by the decomposition of nitric acid from the coalescing affinities of platina and muriatic acid for oxygen; or whether it is composed of a *peculiar* gas, analogous to oxigenated muriatic acid and nitrogen, generated from some unknown affinities." The real nature of the gas, which has also been considered by Lavoisier as a particular species, not hitherto described, was first established by Gay Lussac, when Davy had himself proved that "oxigenated muriatic acid" was a simple substance.

In the second section the combinations and composition of nitrous oxide are investigated, and an account is given of its decomposition by combustible bodies, and a series of experiments are described which are now among the stock illustrations of the chemical lecture-room. As to its composition, he says, "taking the mean estimations from the most accurate experiments, we may conclude that 100 grains of the known ponderable matter of nitrous oxide consist of about 36·7 oxygen and 63.3 nitrogen"—no very great disparity from modern numbers, *viz.* 36·4 oxygen and 63·6 nitrogen. He concludes this section with a short review of the characteristic properties of the combinations of oxygen and nitrogen, among which he is led to class atmospheric air.

> That the Oxygen and nitrogen of atmospheric air exist in chemical union, appears almost demonstrable from the following evidences:
>
>   1st: The equable diffusion of Oxygen and nitrogen through every part of the atmosphere, which can hardly be supposed to depend on any other cause than an affinity between these principles.
>
>   2dly: The difference between the specific gravity of atmospheric air, and a mixture of 27 parts oxygen and 73 nitrogen, as found by calculation; a difference apparently owing to expansion in consequence of combination.

These "evidences" had already been adduced by others, as stated by Davy; the first was subsequently disproved by Dalton, the second was based on inaccurate analyses of air.

To these Davy added two other "proofs" which originated with him:

>   3dly: The conversion of nitrous oxide into nitrous acid, and a gas analogous to common air, by ignition.
>
>   4thly: The solubility of atmospheric air undecompounded.

Of these it may be stated that the first is invalid, and the second not true. Nitrous oxide may, under certain circumstances, give rise to a mixture of oxygen and nitrogen, but not necessarily in the same proportion as in common air; and the air boiled out from water has not the same composition as atmospheric air.

Davy a few years afterwards obtained much clearer views as to the real nature of the atmosphere, and was, in fact, one of the earliest to recognise that it is merely a mixture of oxygen and nitrogen.

The third section consists of an account of observations on the action of nitrous oxide upon animals, and an investigation of the changes effected in it by respiration; whilst the fourth and last gives a history of the respirability and of the extraordinary effects of nitrous oxide, with details of experiments on its powers made by different individuals. The last portion of the inquiry— in time of execution the first—is particularly interesting to the biographer of Davy, not only because the work in it was originated and carried out by him, but also from the light it incidentally throws on his character and genius:

"A short time," he says, "after I began the study of chemistry, in March 1798, my attention was directed to the dephlogisticated nitrous gas of Priestley, by Dr Mitchell's Theory of Contagion." "Dr Mitchell," he tells us in a foot-note, "attempted to prove from some phenomenon connected with contagious diseases, that dephlogisticated nitrous gas which he called oxide of septon, was the principle of contagion, and capable of producing the most terrible effects when respired by animals in the minutest quantities, or even when applied to the skin or muscular fibre." "The fallacy of this theory," he continues, "was soon demonstrated by a few coarse experiments made on small quantities of the gas procured from zinc and diluted nitrous [nitric] cid. Wounds were exposed to its action, the bodies of animals were immersed in it without injury; and I breathed it mingled in small quantities with common air, without remarkable effects. An inability to procure it in sufficient quantities prevented me at this time from pursuing the experiments to any greater extent. I communicated an account of them to Dr Beddoes."

In the early part of April, 1799, he obtained nitrous oxide in a state of purity, and, as already stated, made the attempt to respire it.

"I was aware," he says, "of the danger of this experiment. It certainly would never have been made if the hypothesis of Dr Mitchell had in the least influenced my mind. I thought that the effects might be possibly depressing and painful, but there were many reasons which induced me to believe that a single inspiration of a gas apparently possessing no immediate action on the irritable fibre, could neither destroy nor immediately injure the Powers of life."

The experiment was made: the gas passed into the bronchia without stimulating the glottis, and produced no uneasy feeling in the lungs. There was a sense of fullness in the head accompanied with loss of distinct sensation, and voluntary power, a feeling analogous to that produced in the first stage of intoxication but unattended by pleasurable sensation. In company with Dr Beddoes the experiment was repeated, with the following results:

Having previously closed my nostrils and exhausted my lungs, I breathed four quarts of nitrous oxide from and in to a silk bag. The first feelings were similar to those produced in the last experiment, but in less than half a minute, the respiration being continued, they diminished gradually, and were succeeded by a sensation analogous to gentle pressure on all the muscles attended by a highly pleasurable thrilling, particularly in the chest and the extremities. The objects around me became dazzling, and my hearing more acute. Towards the last inspirations, the thrilling increased, the sense of muscular power became greater, and at last an irresistible propensity to action was indulged in; I recollect but indistinctly what followed; I know that my motions were various and violent. These effects very soon ceased after respiration. In ten minutes I had recovered my natural state of mind. The thrilling in the extremities continued longer than the other sensations. This experiment was made in the morning; no langour or exhaustion was consequent, my feelings throughout the day were as usual, and I passed the night in undisturbed repose. The next morning the recollections of the effects of the gas were very indistinct, and had not remarks written immediately after the experiment recalled them to my mind I should have even doubted of their reality. I was willing indeed to attribute some of the strong emotion to the enthusiasm, which I supposed must have been necessarily connected with the perception of agreeable feelings, when I was prepared to experience painful sensations. Two experiments, however, made in the course of this day, with scepticism, convinced me that the effects were solely owing to the specific operation of the gas.

Having thus ascertained the powers of the gas, he made many experiments to ascertain the length of time it might be breathed with safety, its action on the pulse, and its general effects on the health when often respired.

After a number of experiments made to determine its effect in allaying fatigue, in inducing sleep, or in alleviating the after-effects of vinous intoxication, he resolved

to breathe the gas for such a time and in such quantities as to produce excitement equal in duration and superior in intensity to that occasioned by high intoxication from opium or alcohol.

For this purpose he was enclosed in an air-tight or box-chamber, into which from time to time, by the help of Dr Kinglake, successive quantities of twenty quarts of nitrous oxide were introduced. As he breathed the gas, he found that his temperature and pulse gradually increased. He experienced a generally diffused warmth without the slightest moisture of the skin, a sense of exhilaration similar to that produced by a small dose of wine, and disposition to muscular motion and to merriment. Luminous points seemed frequently to pass before his eyes, his hearing became more acute, and he felt a pleasant lightness and power of exertion in the muscles; and, on account of the great desire of action, rest was painful. After having been in the box for an hour and a quarter he began to respire twenty quarts of unmingled nitrous oxide. What followed is best described in his own words:

> A thrilling, extending from the chest to the extremities, was almost immediately produced. I felt a sense of tangible extension highly pleasurable in every limb; my visible impressions were dazzling, and apparently magnified, I heard distinctly every sound in the room, and was perfectly aware of my situation. By degrees, as the pleasurable sensations increased, I lost all connection with external things; trains of vivid visible images rapidly passed through my mind, and were connected with words in such a manner, as to produce perceptions perfectly novel. I existed in a world of newly connected and newly modified ideas I theorised, I imagined that I made discoveries. When I was awakened fro this semi-delirious trance by Dr Kinglake, who took the bag from my mouth, indignation and pride were the first feelings produced by the sight of the persons about me. My emotions were enthusiastic and sublime, and for a minute I walked round the room perfectly regardless of what was said to me. As I recovered my former state of mind I felt an inclination to communicate the discoveries I had made during the experiment. I endeavoured to recall the ideas: they were feeble and indistinct; one collection of terms however presented itself ; and with a most intense belief and prophetic manner, I exclaimed to Dr Kinglake, *'Nothing exists but thoughts! The universe is composed of impressions, ideas, pleasures and pains!'*

As might be anticipated, the friend of Coleridge and Southey, himself a youth of sensibility and poetic feeling, was curious to learn whether this wonderful gas would increase his stock of the divine afflatus. He walked amidst the scenery of the Avon, "rendered exquisitely beautiful by bright moonshine," and, with a mind filled with pleasurable feelings, he breathed the gas, and we have as a consequence the following effusion:

> Not in the ideal dreams of wild desire
> Have I beheld a rapture-wakening form
> My bosom burns with no unhallow'd fire,
> Yet is my cheek with rosy blushes warm;
> Yet are my eyes with sparkling lustre fill'd;
> Yet is my mouth replete with murmuring sound
> Yet are my limbs with inward transports fill'd,
> And clad with new-born mightiness around.

Whether, as the result of this effort, Davy ever again essayed to tempt the muse when under the influence of nitrous oxide is doubtful. Nowadays the gas is too frequently associated with unhappy memories of the dentist's chair to call up pleasurable associations in a poet's mind.

Davy concludes his memoir with some cautious speculations as to the mode of action of nitrous oxide. That it acts on the blood he was well aware, but it has been left for subsequent research to determine in what manner. He points out that "as nitrous oxide in its extensive operation appears capable of destroying physical pain, it may probably be used with advantage during surgical operations in which no great effusion of blood takes place." As is well known, nitrous oxide is now one of the commonest ancesthetic agents.

As regards the general question how far the gases are likely to subserve the interests of medicine, he is very guarded.

"Pneumatic chemistry," he says, "in its application to medicine is an art in infancy, weak, almost useless, but apparently possessed of capabilities of improvement. To be rendered strong and mature, she must be nourished by facts, strengthened by exercise, and cautiously directed in the application of her powers by rational scepticism."

Davy's success with nitrous oxide led him to attempt to respire other gases—such as hydrogen, nitric oxide, carbonic acid—with in one or two cases almost fatal consequences. On one occasion he tried to breathe water-gas, made by passing steam over charcoal, and was with difficulty brought to life again. These deleterious experiments, carried on with all the ardour and impetuosity of his nature, and at the expense of much nervous energy, reacted prejudicially on his health, and he was obliged to seek relaxation and quiet in the pure atmosphere of his native place.

With the approach of winter he was back again in Bristol, with health restored and vigour renewed. The following letter to Mr Davies Gilbert is interesting as fixing the time at which he entered on the path of inquiry which was to lead him to his greatest triumphs:

Pneumatic Institution, *Oct.* 20, 1800.

In pursuing experiments on galvanism, during the last two months, I have met with unexpected and unhoped-for success. Some of the new facts on this subject promise to afford instruments capable of destroying the mysterious veil which Nature has thrown over the operations and properties of ethereal fluids.

Galvanism I have found, by numerous experiments, to be *a process purely chemical*, and to depend wholly on the oxidation of metallic surfaces, having different degrees of electric conducting power.

Zinc is incapable of decomposing *pure* water; and if the zinc plates be kept moist with *pure* water, the galvanic pile does not act; but zinc is capable of oxidating itself when placed in contact with water, holding in solution either oxygen, atmospheric air, or nitrous or muriatic acid, &c; and under such circumstances the galvanic phenomena are produced, and their intensity is in proportion to the rapidity with which the zinc is oxidated.

The galvanic pile only acts for a few minutes, when introduced into hydrogen, nitrogen, or hydro-carbonate [the gas obtained by the action of steam on charcoal]; that is, only as long as the water between its plates holds some oxygen in solution; immerse it for a few moments in water containing air, and it acts again.

It acts very vividly in oxygen gas, and less so in the atmosphere. When its plates are moistened by marine acid, its action is very powerful, but infinitely more so when nitrous [nitric] acid is employed. Five plates with nitrous [nitric] acid gave sparks equal to those of the common pile. From twenty plates the shock was insupportable.

I had almost forgotten to mention, that charcoal is a good galvanic exciter, and decomposes water, like the metals, in the pile; but I must stop, without being able to expatiate on the connection which is now obvious between galvanism and some of the phenomena of organic motion …

I remain with sincere respect and affection, yours
Humphry Davy.

To his mother he writes:

Hotwells, *November* 19, 1800.

My dear Mother,—Had I believed that my silence of six weeks would have given you a moment's uneasiness, I should have written long ago. But I have been engaged in my favourite pursuit of experimenting, and in endeavouring to amuse two of my friends who have spent some days at the Institute. One of them is your quondam lodger, Gregory Watt, who desired to be kindly remembered to you and the family …

Accept my affectionate thanks for your presents. I have received them all, and I have made a good use of them all. Several times has a supper on the excellent marinaded pilchards made me recollect former times, when I sat opposite to you, my dear mother, in the little parlour, round the little table eating of the same delicious food, and talking of future unknown things. Little did I then think of my present situation, or of the mode in which I am, and am to be, connected with the world. Little did I then think that I should ever be so long absent from the place of my birth as to feel longings so powerful as those I now feel for visiting it again ...

I shall see with heartfelt pleasure the time approaching when I shall again behold my first home—when I shall endeavour to repay some of the debts of gratitude I owe to you, to the Doctor [Tonkin] and to my aunts. My next visit shall not be so short a one as the last. I will stay with you at least two or three months. You have let half your house. Have you a bed-room reserved for me, and a little room for a laboratory? Which part have you let? When I come to Penzance we will settle all about John till then I should like for him to learn French and Latin with Mr Dugart. The expense of this or any other part of his education I will be glad to defray. Do not by any means put him with Mr Coryton ... I will write to Kitty in the course of next month. I am glad to hear Grace is better ...

All in the way of progress goes on nobly. My health was never better than it has been since I left Cornwall last. I shall be very glad to hear from you soon. You have a hundred objects to write about interesting to me. I can only write of myself ...

Love to Kitty, Grace, Betsy and John.

Farewell, my dear mother. I am your affectionate son

H. Davy.

The following letter is to his old friend and benefactor, Mr John Tonkin:

Dowry Square, Clifton, *Jan.* 12, 1801.

Respected Sir, ... Natural philosophy has lately been enriched with many curious discoveries, amongst which galvanism, a phenomenon that promises to unfold to us some of the laws of our nature is one of the most important. In medicine, the inoculation for the cow-pox is becoming general, not in England alone, but over the whole of Europe; and taking circumstances as they now stand, it promises gradually to annihilate small-pox. My discoveries relating to the nitrous oxide, the pleasure-producing air, are beginning to make some noise; the experiments have been repeated, with the greatest success, by the professors of the University of Edinburgh, who have taken up the subject with great ardour; and I have received letters of thanks and of praises for my labours from some of the most respectable of the English philosophers. I am sorry to be so much of an egotist; yet I cannot speak of the Pneumatic Institution and its success without speaking of myself. Our patients are becoming daily more numerous, and our Institution, in spite of the political odium attached to its founder, is respected, even in the trading city of Bristol ... I am at this moment very healthy and very happy; I have had great success in my experiments and I gain a competence by my pursuits, at the same time that I am (in hopes at least) doing something towards promoting the public good. If I feel any anxiety, it is that of being removed so far from you, my mother, and my relations and friends. If I was nearer, I would endeavour to be useful to you; I would endeavour to pay some of the debts of gratitude I owe to you, my first protector and earliest friend. As it is, I must look forward to a futurity that will enable me to do this; but, believe me, wherever I am, and whatever may be my situation, I shall never lose the remembrance of obligations conferred on me, or the sense of gratitude which ought to accompany them.

> I remain, respected Sir, with unfeigned duty and affection, yours
> H. DAVY.

1.　*Cf.* An account of several veins of Sulphate of Strontites, found in the neighbourhood of Bristol, with an Analysis of the different varieties. By W. Clayfield. "Nicholson's Journ." III, 1800, pp. 36-41

III

THE PNEUMATIC INSTITUTION, BRISTOL: 1798–1801 (CONTINUED)

PERHAPS AT NO TIME OF his life was Davy more keenly sensible of the joy of living than at this period—"in the flower and freshness of his youth," as Southey says. That he was eager, active, buoyant, happy, is obvious from his letters. He had the sweet consciousness of success, and all the sweeter that it had so quickly followed the bitterness of disappointment. He had been able to measure himself against some of the ablest minds of the time—of men who were making the intellectual history of the early part of this century—and the comparison, we may be sure, was not altogether unpleasiug to him.

The love of fame—"the honourable need of the applause of enlightened men," as he called it—was his ruling passion and the motive principle of his life. As his experience and the range of his knowledge widened, he felt a growing conviction that with health and strength he need set no bounds to the limits of his ambition.

Of the impression he made on others, and of the influence and power he exerted on minds far more matured than his own, we have abundant evidence in the letters of his contemporaries. Miss Edgeworth's good-humoured patronage quickly passed into amazement and ended in awe. Writing to William Taylor of Norwich, Southey calls Davy "a miraculous young man, whose talents I can only wonder at." Amos Cottle, poet and publisher, to whom he was introduced shortly after his arrival at Bristol, says of him in the *Reminiscences of Coleridge and Southey*:

I was much struck with the intellectual character of his face. His eye was piercing, and when not engaged in converse, was remarkably introverted, amounting to absence, as though his mind had been pursuing some severe train of thought scarcely to be interrupted by external objects; and, from the first interview also, his ingenuousness impressed me as much as his mental superiority.

Cottle on one occasion said to Coleridge, "During your stay in London you doubtless saw a great many of what are called the cleverest men—how do you estimate Davy in comparison with these?" Mr Coleridge's reply was strong but expressive: "Why, Davy can eat them all! There is an energy, an elasticity, in his mind which enables him to seize on and analyse all questions, pushing them to their legitimate consequences. Every subject in Davy's mind has the principle of vitality. Living thoughts spring up like turf under his feet." It can hardly be doubted that Davy's connection with that remarkable literary coterie which made its headquarters in the neighbourhood of Bristol in the last year of the eighteenth century, strongly stimulated his intellectual activity. In one of his poems written at this period he speaks of having

felt the warmth,
The gentle influence of congenial souls,
Whose kindred hopes have cheer'd me

That these "congenial souls" in turn felt his no less strongly will be apparent from the letters—the first from Southey, who then Westbury, the others from Coleridge, who removed to the Lake country:

Thursday, *May 4th*, 1799.

Your 'Mount's Bay,' my dear Davy, disappointed me in its length. I expected more, and wished more, because what there is is good; there is a certain swell, an elevation in the flow of the blank verse, which, I do not know how, produces an effect like the fullness of an organ-swell upon the feeling. I have felt it from the rhythm of Milton, and sometimes of Akenside, a pleasure wholly independent from that derived from the soul of the poetry, arising from the beauty of the body only. I believe a man who did not understand a word of it would feel pleasure and emotion at hearing such lines read with the tone of a poet ...

I must not press the subject of poetry upon you, only do not lose the feeling and the habit of seeing all things with a poet's eye; at Bristol you have a good society, but not a man who knows anything of poetry. Dr Beddoes' taste is very pessimism. Cottle only likes what his friends and himself write. Every person fancies himself competent to pronounce upon the merits of a poem, and yet no trade requires so long an apprenticeship, or involves the necessity of such multifarious knowledge …

At Lymouth I saw Tobin's friend Williams who opened upon me with an account of the gaseous oxide. I had the advantage of him, having felt what he it seems had only seen. Lynmouth where he is fixed is certainly the most beautiful place I have seen in England, so beautiful that all the after-scenes come flat and uninteresting. The Valley of Stones is about half a mile distant, a strange and magnificent place, which ought to have filled the whole neighbourhood with traditions of giants, devils, and magicians, but I could find none, not even a lie preserved. I know too little of natural history to hypothesize upon the cause of this valley; it appeared to me that nothing but water could have so defleshed and laid bare the bones of the earth—that any inundation which could have overtopped these heights must have deluged the kingdom; but the opposite hills are clothed with vegetable soil and verdure, therefore the cause must have been partial—a waterspout might have occasioned it perhaps—and there my conjectures rested, or rather took a new direction to the pre-Adamite kings, the fiends who married Diocletian's fifty daughters—their giant progeny, old Merlin and the builders of the Giant's Causeway.

For the next Anthology I project a poem on our Clifton rocks ; the scenery is fresh in my sight, and these kind of poems derive a more interesting cast as *recollections* than as immediate pictures. Farewell.

Yours truly,
ROBERT SOUTHEY.

Keswick, Friday Evening, *July* 25, 1800.

MY DEAR DAVY,—Work hard, and if success do not dance up like the bubbles in the salt (with the spirit lamp under it[1]) may the Devil and his dam take success! My dear fellow! from the window before me there is a great *camp* of mountains. Giants seem to have pitched their tents there. Each mountain is a giant's tent, and how the light streams from them! Davy! I *ache* for you to be with us.

W. Wordsworth is such a lazy fellow, that I bemire myself by making promises for him: the moment I received your letter, I wrote to him. He will, I hope, write immediately to Biggs and Cottle. At all events, these poems must not as yet be delivered up to them, because that beautiful poem, 'The Brothers,' which I read to you in Paul Street, I neglected to deliver to you, and that must begin the volume. I trust, however, that I have invoked the sleeping bard with a spell so potent, that he will awake and deliver up that Sword of Argantyr, which is to rive the enchanter *Gaudyverse* from his crown to his feet.

What did you think of that case I translated for you from the German? That I was a well-meaning sutor who had ultra-crepidated with mere zeal than wisdom!! I give myself credit for that word 'ultra-crepidated,' it started up in my brain like a creation ...

We drank tea the night before I left Grasmere, on the island in that lovely lake; our kettle swung over the fire, hanging from the branch of a fir-tree, and I lay and saw the woods, and mountains, and lake all trembling, and as it were idealized through the subtle smoke, which rose up from the clear red embers of the fir-apples which we had collected; afterwards we made a glorious bonfire on the margin, by some elder bushes, whose twigs heaved and sobbed in the uprushing column of smoke, and the image of the bonfire, and of us that danced round it, ruddy, laughing faces in the twilight; the image of this in a lake, smooth as that sea, to whose waves the Son of God had said, *Peace!* May God, and all his sons, love you as I do.

S.T. COLERIDGE.

Sara desires her kind remembrances. Hartley is a spirit that dances on an aspen leaf: the air that yonder sallow-faced and yawning tourist is breathing, is to my babe a perpetual nitrous oxide ...

Thursday night, *Oct.* 9, 1800.

MY DEAR DAVY,—I was right glad, glad with a *stagger* of the heart, to see your writing again. Many a moment have I had all my France and England curiosity suspended and lost, looking in the advertisement front column of the Morning Post Gazetteer, for *Mr Davy's Galvanic habitudes of charcoal.* Upon my soul, I believe there is not a letter in those words round which a world of imagery does not circumvolve; your room, the garden, the cold bath, the moonlit rocks ... and dreams of wonderful things attached to your name ... I pray you do write to me immediately, and tell me what you mean by the possibility of your assuming a new occupation; have you been successful to the extent of your expectations in your late chemical inquiries?

As to myself, I am doing little worthy the relation. I write for Stuart in the Morning Post, and I am compelled by the god Pecunia, which was one name of the supreme Jupiter, to give a volume of letters from Germany, which will be a decent *lounge* book, and not an atom more. The Christabel was running up to 1,300 lines, and was so much admired by Wordsworth, that he thought it indelicate to print two volumes with his name, in which so much of another man's was included ... We mean to publish the Christabel, therefore, with a long blank-verse of Wordsworth's, entitled 'The Pedlar' [afterwards changed to 'The Excursion']. I assure you I think very differently of Christabel. I would rather have written Ruth and Nature's Lady, than a million such poems. But why do I calumniate my own spirit by saying I would rather? God knows it is as delightful to me that they *are* written ...

Wordsworth is fearful you have been much teazed by the printers on his account, but you can sympathise with him ...

When you write, and do write soon, tell me how I can get your Essay on the Nitrous Oxide ... Are your galvanic discoveries important? What do they lead to? All this is *ultra-crepidation*, but would to heaven I had as much knowledge as I have sympathy! ...

God bless you! Your most affectionate
S.T. COLERIDGE.

Greta Hall, Tuesday night, *Dec.* 2, 1800.
MY DEAR DAVY,—By an accident I did not receive your letter till this evening. I would that you had added to the account of your indisposition the probable causes of it. It has left me anxious whether or no you have not exposed yourself to unwholesome influences in your chemical pursuits. There are *few* beings both of hope and performance, but few who combine the 'are' and the 'will be.' For God's sake, therefore, my dear fellow, do not rip open the bird that lays the golden eggs ...

At times, indeed, I would fain be somewhat of a more tangible utility than I am; but so I suppose it is with all of us—one while cheerful, stirring, feeling in resistance nothing but a joy and a stimulus; another while drowsy, self-distrusting, prone to rest, loathing our own self-promises, withering our own hopes—our hopes, the vitality and cohesion of our being?

I purpose to have Christabel published by itself—this I publish with confidence—but my travels in Germany come from me now with mortal pangs ...

Wordsworth has nearly finished the concluding poem. It is of a mild, unimposing character, but full of beauties to those short-necked men who have their hearts sufficiently near their heads—the relative distance of which (according to citizen Tourder, the French translator of Spallanzani) determines the sagacity or stupidity of all bipeds and quadrupeds ...

God love you!
S.T. COLERIDGE.

"No man ever had genius who did not aim to execute more than he was able." So wrote Davy in one of his early notebooks; and of no man was this more true than of Davy himself. Busy as he was with experimental research at this time, his mind was by no means wholly occupied with it. Change of mental occupation was, indeed, a necessity to him. At no period of his life could he exercise that power of sustained and concentrated thought which so strikingly characterised Newton or Dalton or Faraday. The following scheme of intellectual work which he marked out for himself shortly after his arrival in Bristol, is characteristic of the restless, changeful activity of his mind

"*Resolution:* To work two hours with pen before breakfast on the 'Lover of Nature'; and 'The Feelings of Eldon' from six till eight; from nine till two in experiments; from four to six, reading; seven till ten, metaphysical reading (*i.e.* 'System of the Universe')." The "Lover of Nature" and "The Feelings of Eldon" were two among the half-dozen romances he projected at one time or other, and of which fragments were found amongst his papers, and by means of which he intended to inculcate his own metaphysical and philosophical ideas and his views on education and the development of character. Dr John Davy tells us that his notebooks at this period were not less characteristic; "they contain, mixed together, without the least regard to order, schemes and minutes of experiments, passing thoughts of various kinds, lines of poetry (but these are in small proportion), fragments of stories and romances, metaphysical fragments, and sketches of philosophical essays."

Many of these jottings and reflections are evidently based on his own experience, and hence serve to illustrate his temperament and the workings of his mind. In an essay on "Genius," written at this time, he says:

Great powers have never been exerted independent of strong feelings. The rapid arrangements of ideas, from their various analogies to the equally rapid comparisons of these analogies, with facts uniformly occurring during the progress of discovery, have existed only in those minds where the agency of strong and various motives is perceived—of motives modifying each other, mingling with each other, and producing that fever of emotion, which is the joy of existence and the consciousness of life.

The following extracts relate to science and philosophy:

Philosophy is simple and intelligible. We owe confused systems to men of vague and obscure ideas.

We ought to reason from effects alone. False philosophy has uniformly depended upon making use of words which signify no definite ideas.

Experimental science hardly ever affords us more than approximations to truth; and whenever many agents are concerned we are in great danger of being mistaken.

Scepticism in regard to theory is what we ought most rigorously to adhere to.

The feeling generally connected with new facts enables us to reason more rapidly upon them, and is peculiarly active in calling up analogies.

Probabilities are the most we can hope for in our generalisation, and whenever we can trace the connection of a series of facts, without being obliged to imagine certain relations, we may esteem ourselves fortunate in our approximations.

One use of physical science is that it gives definite ideas.

To the same period belongs the sketch or plan of a poem, in blank verse, in six books, on the deliverance of the Israelites from Egypt, which either Southey or Coleridge had proposed to him as a joint-work, fragments of which are to be found amongst the notebooks.

Towards the end of 1800 Davy's visions of future greatness began to take more definite shape. This is hinted at in the letter from Coleridge of October 9th, 1800, already given, and also in one to his mother, dated September 27th, 1800, in which he says, "My future prospects are of a very brilliant nature, and they have become more brilliant since I last wrote to you; but wherever there is uncertainty I shall refrain from anticipating."

In a few months the uncertainty was practically at an end.

He had been drawn into the great vortex called London, "full," as he says in a letter to Hope, "of the expectation of scientific discovery from the action of mind upon mind in this great hot-bed of human power." He thus informs his mother:

31st January, 1801.

MY DEAR MOTHER,—During the last three weeks I have been very much occupied by business of a very serious nature. This has prevented me from writing to you, to my aunt, and to Kitty. I now catch a few moments only of leisure to inform you that I am exceedingly well, and that I have had proposals of a very flattering nature to induce me to leave the Pneumatic Institution for a permanent establishment in London.

You have perhaps heard of the Royal Philosophical Institution, established by Count Rumford, and others of the aristocracy. It is a very splendid establishment, and wants only a combination of talents to render it eminently useful.

Count Rumford has made proposals to me to settle myself there, with the present appointment of assistant lecturer on chemistry, and experimenter to the Institute; but this only to prepare the way for my being in a short time sole professor of chemistry, &c.; an appointment as honourable as any scientific appointment in the kingdom, with an income of at least £500 a year.

I write today to get the specific terms of the present appointment, when I shall determine whether I shall accept of it or not. Dr Beddoes has honourably absolved me from all engagements at the Pneumatic Institution, provided I choose to quit it. However, I have views here which I am loath to leave, unless for very great advantages.

You will all, I dare say, be glad to see me getting amongst the *Royalists,* but I will accept of no appointment except upon the sacred terms of *independence* ...

> I am your most affectionate son
> H. DAVY.

In the middle of February he was in London negotiating with Rumford. He wrote to his mother, "His proposals have not been unfair, and I have nearly settled the business." How the business was actually settled appears from the following extract from the Minute Book of the Royal Institution of a resolution adopted at a Meeting of the Managers on February 16th, 1801:

Resolved—That Mr Humphry Davy be engaged in the service of the Royal Institution, in the capacities of Assistant Lecturer in Chemistry, Director of the Laboratory, and Assistant Editor of the Journals of the Institution, and that he be allowed to occupy a room in the house, and be furnished with coals and candles; and that he be paid a salary of one hundred guineas per annum.

He returned to Bristol to hand over his charge of the Pneumatic Institution, and to take leave of his many friends in that city. The following letter to Mr Davies Gilbert is interesting and characteristic:

> Hotwells, *March 8th*, 1801.

I cannot think of quitting the Pneumatic Institution, without giving you intimation of it in a letter; indeed, I believe I should have done this some time ago, had not the hurry of business, and the fever of emotion produced by the prospect of novel changes in futurity, destroyed to a certain extent my powers of consistent action.

You, my dear Sir, have behaved to me with great kindness, and the little ability I possess you have very much contributed to develope; I should therefore accuse myself of ingratitude were I to neglect to ask your approbation of the measures I have adopted with regard to the change of my situation, and the enlargement of my views in life.

In consequence of an invitation from Count Rumford, given to me with some proposals relative to the Royal Institution, I visited London in the middle of February, where, after several conferences with that gentleman, I was invited by the Managers of the Royal Institution to become the Director of their laboratory, and their Assistant Professor of Chemistry; at the same time I was assured that, within the space of two or three seasons, I should be made sole Professor of Chemistry, still continuing Director of the laboratory.

The immediate emolument offered was sufficient for my wants; and the sole and uncontrolled use of the apparatus of the Institution, for private experiments, was to be granted me. The behaviour of Count Rumford, Sir Joseph Banks, Mr Cavendish, and the other principal managers, was liberal and polite; and they promised me any apparatus that I might need for new experiments.

The time required to be devoted to the services of the Institution was but short, being limited chiefly to the winter and spring. The emoluments to be attached to the office of sole Professor of Chemistry are great; and, above all, the situation is permanent, and held very honourable.

These motives, joined to the approbation of Dr Beddoes, who with great liberality has absolved me from my engagements at the Pneumatic Institution, and the strong wishes of most of my friends in London and Bristol, determined my conduct.

Thus I am quickly to be transferred to London, whilst my sphere of action is considerably enlarged, and as much power as I could reasonably expect, or even wish for at my time of life, secured to me without the obligation of labouring at a profession.

The Royal Institution will, I hope, be of some utility to Society. It has undoubtedly the capability of becoming a great instrument of moral and intellectual improvement. Its funds are very great. It has attached to it the feelings of a great number of people of fashion and property, and consequently may be the means of employing, to useful purposes, money which would otherwise be squandered in luxury, and in the production of unnecessary labour. Count Rumford professes that it will be kept distinct from party politics; I sincerely wish that such may be the case, though I fear it. As for myself, I shall become

attached to it full of hope, with the resolution of employing all my feeble powers towards promoting its true interests.

So much of my paper has been given to pure egotism, that I have but little
room left to say anything concerning the state of science …

Here, at the Pneumatic Institution, the nitrous oxide has evidently been of
use. Dr Beddoes is proceeding in the execution of his great popular physiologi-
cal work, which, if it equals the plan he holds out, ought to supersede every
work of the kind.

I have been pursuing Galvanism with labour, and some success. I have been
able to produce galvanic power from simple plates, by effecting on them different oxidating and de-oxidating processes; but on this point I cannot enlarge in
the small remaining space of paper …

It will give me sincere pleasure to hear from yon, when you are at leisure.
After the 11th I shall be in town—my direction, Royal Institution, Albemarle
Street. I am, my dear friend, with respect and affection,

Yours,
HUMPHRY DAVY.

With Davy's departure we, too, may take our leave of the Pneumatic
Institution. Like most of Dr Beddoes's performances, it—to use Davy's
words—failed to equal the plan its projector held out. It struggled on for
awhile, living on such success as Davy had brought it, and ultimately died of
inanition. Its founder ended his days a disappointed man, and on his deathbed
wrote to his former assistant, in connection with whom his memory mainly
lives, "like one who has scattered abroad the *Avena fatua* of knowledge, from
which neither branch, nor blossom, nor fruit, has resulted, I require the consolations of a friend."

1.   Doubtless an allusion to the decomposition of ammonium nitrate, which Coleridge had
frequently seen Davy effect.

IV
THE ROYAL INSTITUTION

THE ROYAL INSTITUTION, AS ORIGINALLY conceived, was an establishment for the benefit of the poor. It was founded at the close of the last century by Benjamin Thomson, a Royalist American in the service of the Elector Palatine of Bavaria, by whom he was created a Count of the Holy Roman Empire. Count Rumford, as he is commonly called, was a practical philanthropist and a man of science, best known to this age by his association with the present-day doctrine of the nature of heat; and to his contemporaries, by his constant efforts to apply science to domestic economy. In 1796 Rumford put forth a "proposal for forming in London by private subscription an establishment for feeding the poor, and giving them useful employment, and also for furnishing food at a cheap rate to others who may stand in need of such assistance, connected with an institution for introducing and bringing forward into general use new inventions and improvements, particularly such as relate to the management of heat and the saving of fuel, and to various other mechanical contrivances by which domestic comfort and economy may be promoted." Rumford, as he says in one of his letters to Thomas Bernard—another practical philanthropist, and one of his earliest associates in the undertaking here referred to—was deeply impressed with the necessity of rendering it *fashionable* to care for the poor and indigent." The immediate result was the foundation of the Society for Bettering the Condition of the Poor; but as regards the associated Institution, it was eventually considered that it would be "too conspicuous, and too interesting and important, to be made *an appendix* to any other existing establishment, and consequently it must stand alone, and on its own proper basis."

In 1799, Rumford conferred with the Committee of the Society for Bettering the Condition of the Poor as to the steps to be taken to found, "by private subscription, a public institution for diffusing the knowledge and facilitating the general and speedy introduction of new and useful mechanical inventions and improvements; and also for teaching, by regular courses of philosophical lectures and experiments, the applications of the new discoveries in science to the improvement of arts and manufactures, and in facilitating the means of procuring the comforts and conveniences of life." The Institution was duly launched in March, 1799, with Sir Joseph Banks as Chairman of Managers, Count Rumford as Secretary, and Mr Thomas Bernard, the promoter of the Institution for the Protection and Instruction of Climbing Boys, and of the Society for the Relief of Poor Neighbours in Distress, as Treasurer. The second volume of the "Reports of the Society for Bettering the Condition of the Poor" contains a long account of the Institution, "so far as it may be expected to affect the poor," from the pen of Mr Bernard, concerning which Dr Bence Jones, a former Secretary of the Institution, dryly remarks, "It is difficult to believe that the Royal Institution of the present day was ever intended to resemble the picture given of it in this Report."

Rumford, from the outset, threw himself with great zeal and ardour into the work of organising and starting the Institution, and it was mainly by his energy and administrative ability that so speedy a beginning was made. Mr Mellish's house in Albemarle Street was bought, and its apartments were quickly transformed into lecture rooms, model rooms, library, offices, etc. In May "a good cook was engaged for the improvement of culinary advancement—one object, and not the least important—for the Royal Institution." Rumford was requested by the Managers to live in the house, to superintend the servants, to preserve order and decorum, and to control the expenses of housekeeping.

Towards the end of 1799 Dr Garnett was secured as Lecturer and Scientific Secretary. Thomas Garnett, a physician, who at one time practised at Harrogate, and who is known to chemists for his researches into the composition of the Harrogate mineral waters, was at the time Professor of Chemistry and Experimental Philosophy at Anderson's Institution in Glasgow. He had a considerable reputation as a lecturer, on the strength of which he was invited by Rumford to come to London. Garnett's lectures began in March, 1800, in what is now the upper Library of the Institution, and which had been fitted up to accommodate the greatest possible number of auditors "with a greater deference to their curiosity than to their convenience."

Although not altogether unsuccessful at the Institution, Garnett—in spite of "the Northern accent which he still retained in a slight degree, and which rendered his voice somewhat inharmonious to a London audience"—was hardly the type of man required for such a place, and differences soon arose between him and Rumford. To add to his difficulties he had, just prior to his removal from Glasgow, lost his wife, and the event seems to have wholly unnerved him. He grew listless and melancholy; and eventually, in 1801, he was called upon to resign. After leaving the Institution, he struggled on for a time, giving courses of scientific lectures in his own house, and at Tom's Coffee-House in the City, and seeking for practice as a physician. Sick in mind and weak in body, he soon broke down, and died in 1802, at the age of thirty-six, leaving his children penniless. The Managers so far bettered the condition of the poor as to subscribe, on behalf of the Institution, £50 towards the publication of his posthumous work on the "Laws of Animal Life," and to allow the book to be dedicated to them.

Accounts differ as to the precise means by which Davy was brought to the notice of Count Rumford, nor is it very important to know whether it was through the intervention of Davies Gilbert, or Dr Hope, or Mr Underwood, or, as was most probably the case, of all three.

In a letter to Hope now before me Davy writes:

> I believe it is in a great measure owing to your kind mention of me to Count Rumford, that I occupy my present situation in the Royal Institution. I ought to be very thankful to you; for most of my wishes through life are accomplished, as I am enabled to pursue my favourite study, and at the same time to be of some little utility to Society.

This much, at least, is certain: there was an absolute agreement among those who had the best means of judging that no better appointment was possible. And yet, if we are to credit Dr Paris, the first impression produced on Rumford by Davy's personal appearance was highly unfavourable, and the Count would not allow him to lecture in the theatre until he had given a specimen of his abilities in the smaller lecture-room, which old *habitués* of the Royal Institution well remember. Dr Paris adds that his first lecture entirely removed every prejudice, and at its conclusion Rumford exclaimed, "Let him command any arrangements which the Institution can afford." And he was accordingly on the next day promoted to the theatre.

Six weeks after his arrival, he gave his first public lecture. How he acquitted himself, may be gleaned from the following account, given under the heading of the "Royal Institution of Great Britain" in the *Philosophical Magazine*, vol. x., p. 281 (1801):

It must give pleasure to our readers to learn that this new and useful institution, the object of which is the application of science to the common purposes of life, may be now considered as settled on a firm basis …

We have also to notice a course of lectures, just commenced at the institution, on a new branch of philosophy—we mean the Galvanic Phenomena. On this interesting branch Mr Davy (late of Bristol) gave the first lecture on the 25th of April. He began with the history of Galvanism, detailed the successive discoveries, and described the different methods of accumulating galvanic influence … He showed the effects of galvanism on the legs of frogs, and exhibited some interesting experiments on the galvanic effects on the solutions of metals in acids …

Sir Joseph Banks, Count Rumford, and other distinguished philosophers were present. The audience were highly gratified, and testified their satisfaction by general applause. Mr Davy, who appears to be very young, acquitted himself admirably well; from the sparkling intelligence of his eye, his animated manner, and the *tout ensemble*, we have no doubt of his attaining a distinguished eminence.

The Managers were so far satisfied, that at a meeting held on June 1st they passed the following resolutions:

Resolved—That Mr Humphry Davy, Director of the Chemical Laboratory, and Assistant Lecturer in Chemistry, has, since he has been employed at the Institution, given satisfactory proofs of his talents as a Lecturer.

Resolved—That he be appointed, and in future denominated, Lecturer in Chemistry at the Royal Institution, instead of continuing to occupy the place of Assistant Lecturer, which he has hitherto filled.

In the following July, Dr Young ("Phenomenon Young," as he was called at Cambridge), the great exponent of the Undulatory Theory of Light, was engaged as Professor of Natural Philosophy, Editor of the Journals, and General Superintendent of the House.[1]

At a meeting held in the same month, the Managers

Resolved—That a Course of Lectures on the Chemical Principles of the Art of Tanning be given by Mr Davy. To commence the second of November next; and that respectable persons of the trade, who shall be recommended by Proprietors of the Institution, be admitted to these lectures gratis.

To order a young man of twenty-two, who had probably never seen the inside of tannery, to give an account of the art and mystery of leather-making, would seem to savour somewhat of what Coleridge would style "*ultra-crepidation*," and accordingly the Managers further

> *Resolved*—That Mr Davy have permission to absent himself during the months of July, August, and September for the purpose of making himself more particularly acquainted with the practical part of the business of tanning, in order to prepare himself for giving the above-mentioned course of lectures.

Lectures on "The Chemical Principles of the Process of Tanning Leather, and of the objects that must particularly be had in view in attempts to improve that most useful art" are mentioned in Rumford's first prospectus, and the foregoing resolutions were probably passed in consequence. Davy did a considerable amount of experimental work in connection with these lectures, and the Journal of the Royal Institution contains several short communications from him on the chemistry of the subject, but the main facts he discovered are contained in a memoir read to the Royal Society on February 24th, 1803, and published in the *Philosophical Transactions* of that year, under the title of an "Account of Some Experiments and Observations on the constituent Parts of certain astringent Vegetables; and on their Operation in Tanning."

Although Davy, by his earnestness, his knowledge, his felicity of expression, and by a certain dignity of treatment which seemed to invest even the homeliest subjects with unlooked-for importance, could interest an audience on almost any subject he brought before them, we may be sure that his soul soon sighed for a loftier theme than leather. He found it on the occasion of his lecture of January 21st, 1802, when he delivered the introductory discourse of that session. The date, indeed, is a red-letter day not only in Davy's history but also in that of the Royal Institution. From that time the position of the Institution in the scientific and social world of London would seem to be assured.

Its affairs up to this time had been gradually going from bad to worse. The enthusiasm with which it was started a couple of years back had apparently spent itself, and Rumford, by his *hauteur* and high-handed management, had alienated many powerful friends. The subscriptions, which in 1800 had reached £11,047, had fallen in 1802 to £2,999, whilst the expenses were annually increasing. The outlook was gloomy in the extreme, and everything seemed to portend that the latest scheme for the amelioration of humanity was about to share the too common fate of such projects. The young man of twenty-three, however, changed all this as if by the stroke of a magician's wand. No Prince Fortunatus could have done more.

His theme was not too ambitious; it would be considered even trite and commonplace today, and the man would be very bold or very simple who would now attempt to deal with it in the theatre of the Royal Institution; for this introductory lecture was nothing more than an exordium on the worth of science as an agent in the improvement of society. It was, and was felt to be, however, an *apologia* for the very existence of the Institution. Rumford and his fellow managers would seem to have staked everything on a single throw. Davy's power as a lecturer had been noised abroad, and we may be sure that Coleridge and his other friends did not keep their tongues still. Coleridge, indeed, told the literary world that he assiduously attended Davy's lectures, to increase his stock of metaphors. The youth who had discovered "the pleasure-producing air" was talked about in fashionable circles; and Mr Bernard and the Count used their persuasiveness, and Sir Joseph Banks his social power, to secure for him the most cultured audience in London. If we may credit Dr Paris, other influences, too, were at work. Davy's association with Beddoes had probably gained for him the goodwill of the Tepidarians, even if it did not actually give him the *entrée* to the Society; and these Red Republicans, whose "pious orgies" at Old Slaughter's Coffee-House in St Martin's Lane consisted mainly in libations of tea, vied with the Royalists in their efforts to pave his triumphal way. His success was instant and complete. In a series of lofty and impassioned periods he traced the services of science to humanity; he dwelt upon its dignity and nobility as a pursuit, upon its value as a moral and educational force. The small, spare youth, with his earnestness, his eloquence, his unaffected manner, the play of his mobile features, his speaking eyes—"eyes which," as one of his fair auditors was heard to remark, "were made for something besides poring over crucibles"—held his hearers spellbound as he declaimed such sentences as these:

Individuals influenced by interested motives or false views may check for a time the progress of knowledge;—moral causes may produce a momentary slumber of the public spirit;—the adoption of wild and dangerous theories, by ambitious or deluded men, may throw a temporary opprobrium on literature; but the influence of true philosophy will never be despised; the germs of improvement are sown in minds, even where they are not perceived and sooner or later the springtime of their growth must arrive. In reasoning concerning the future hopes of the human species, we may look forward with confidence to a state of society, in which the different orders and classes of men will contribute more effectually to the support of each other than they have hitherto done. This state, indeed, seems to be approaching fast; for, in consequence of the multiplication of the means of instruction the man of science and the manufacturer are daily becoming more assimilated to each other. The artist, who formerly affected to despise scientific principles, because he was incapable of perceiving the advantages of them, is now so far enlightened as to favour the adoption of new processes in his art, whenever they are evidently connected with the diminution of labour; and the increase of projectors, even to too great an extent, demonstrates the enthusiasm of the public mind in its search after improvement …

The unequal division of property and of labour, the differences of rank and condition amongst mankind, are the sources of power in civilized life—its moving causes, and even its very soul. In considering and hoping that the human species is capable of becoming more enlightened and more happy, we can only expect that the different parts of the great whole of society should be intimately united together, by means of knowledge and the useful arts ; that they should act as the children of one great parent, with one determinate end, so that no power maybe rendered useless—no exertions thrown away.

In this view, we do not look to distant ages, or amuse ourselves with brilliant though delusive dreams, concerning the infinite improveability of man, the annihilation of labour, disease, and even death, but we reason by analogy from simple facts, we consider only a state of human progression arising out of its present condition,—we look for a time that we may reasonably expect—FOR A BRIGHT DAY, OF WHICH WE ALREADY BEHOLD THE DAWN.

Those who may read these sentences will either smile at their seeming archaism, or wonder at the antiquity of their argument; for the lesson which Davy inculcated at the beginning of the century is still at its close dinned into our ears, and practically all the stock reasons urged by latter-day writers and platform speakers on technical education and the abstract value of science are to be found in his lectures. But the circumstances of 1802 were widely different from those of 1896. The birth of the century was a singularly auspicious time for science; and many cultured men who knew nothing of science, yet felt in a dim sort of way that it was destined to be a mighty factor in civilisation. Davy's words struck a sympathetic chord; they served to formulate and define ideas of which all who lived in the spirit of the times and shared in its movement must have been conscious. Speaking to willing and receptive ears, and with every attribute of manner, speech, and interest in his favour, he saw his chance; and with a practical sagacity beyond his years, he seized it.

Davy's triumph is recorded in many contemporary notices, and it lives as one of the traditions of the Royal Institution.

Francis Homer thus records his impressions in his journal, under date March 31st, 1802:

> I have been once to the Royal Institution and heard Davy lecture to a mixed and large assembly of both sexes, to the number perhaps of three hundred or more. It is a curious scene; the reflections it excites are of an ambiguous nature, for the prospect of possible good is mingled with the observation of much actual folly. The audience is assembled by the influence of fashion merely; and fashion and chemistry form a very incongruous union ...
>
> Davy's style of lecturing is much in favour of himself, though not, perhaps, entirely suited to the place; it has rather a little awkwardness, but it is that air which bespeaks real modesty and good sense; he is only awkward because he cannot condescend to assume that theatrical quackery of manner which might have a more imposing effect. This was my impression from his lecture. I have since (April 2nd) met Davy in company, and was much pleased with him; a great softness and propriety of manner, which might be cultivated into elegance; his physiognomy struck me as being superior to what the science of chemistry, on its present plan, can afford exercise for; I fancied to discover in it the lineaments of poetical feeling. (*Memoirs of Homer*, vol. i, p. 182.)

Davy's friend Purkis has left us the following still more glowing account:

> The sensation created by his first [second] course of Lectures at the Institution, and the enthusiastic admiration which they obtained, is at this period scarcely to be imagined. Men of the first rank and talent,—the literary and the scientific, the practical and the theoretical, blue stockings, and women of fashion, the old and the young, all crowded—eagerly crowded the lecture-room. His youth, his simplicity, his natural eloquence, his chemical knowledge, his happy illustrations, and well-conducted experiments, excited universal attention and unbounded applause. Compliments, invitations, and presents were showered upon him in abundance from all quarters; his society was courted by all, and all appeared proud of his acquaintance ... A talented lady, since well-known in the literary world, addressed him anonymously in a poem of considerable length, replete with delicate panegyric and genuine feeling ... It was accompanied with a handsome ornamental appendage for the watch, which he was requested to wear when he delivered his next lecture, as a token of having received the poem and pardoned the freedom of the writer.

The anonymous poem "replete with delicate panegyric and genuine feeling" is before me as I write. It is signed "Fidelissima," and is one of several which the same talented lady addressed to him at different times, and which were found among his papers at his death. Some of them, as sonnets, are of considerable merit, and, had space permitted, are well worthy of reproduction.

The Tepidarians—again on the authority of Dr Paris—were delighted. Sanguine in the success of their child—for so they considered Davy—they purposely appointed their anniversary festival on the day of his anticipated triumph. Their dinner was marked by every demonstration of hilarity, and the day was ended by a masquerade at Ranelagh.

Dr John Davy, it should be said, rather sniffs at the Tepidarians and their "ultra-principles," and doubts if his brother ever belonged to their society. Be this as it may, it is certain that the "Royalists" and the fashionable world into which he was drawn soon influenced Davy's social and political views. Dr Davy, whilst willing enough to appreciate at their proper value his brother's natural and intellectual advantages as contributing to his success, points out that other circumstances connected with the Institution and the period conspired to help him:

> The Royal Institution was a new experiment. Novelty in itself is delightful, especially to people of rank and fortune, who at that time in consequence of the Continent being closed, and owing to the war, must have been delighted to have had opened to them a new and unexpected source of interest, fitted to amuse those who were suffering from *ennui,* and to instruct those who were anxious for instruction. The Royal Institution, moreover, was the creation of a large number of influential persons, both in the higher ranks of society and of science. This alone might have sufficed to render it fashionable, and, if fashionable, popular. The period, morally and politically considered, aided the effect; a time of great political excitement had just terminated; a time of gloom and despondency was then commencing. Whatever diverted the public mind and afforded new objects of contemplation, pure and independent sources of amusement and gratification, must have been very welcome to all reflecting persons, even without taking into account the possible and probable good which might be conferred by the Institution on society, in accordance with the intentions with which it was first established.

Davy thus expressed his own feeling of satisfaction to his mother:

London.

My dear Mother,—I have been very busy in the preparation for my lectures; and for this reason I have not written to you. I delivered my second lecture today, and was very much flattered to find the theatre overflowing at this, as well as at the first. I am almost surprised at the interest taken by so many people of rank, in the progress of chemical philosophy; and I hope I am doing a great deal of good, in being the means of producing and directing the taste for it.

I have been perfectly well since I visited Cornwall; and I enter upon my campaign in high health and spirits. After four months of hard but pleasant labour, I shall again be free!

I hope you are all well. I very often reflect upon the times that are past; and my mind is always filled with gratitude to the Supreme Being, who has made us all happy; and that, in placing us in distant parts, and in different circles, neither our feelings or affections have been disturbed …

I shall be very glad to see you again. I intend in June to pass through Scotland and to visit the Western Isles; but I hope I shall spend a part of the autumn with you.

Pray write to me and give me a little news. Beg Kitty and Grace and Betsy and John to recollect me.

I am, my dear mother, your very affectionate son
H. Davy.

The interest and spirit of enthusiasm thus roused was sedulously cultivated by Davy, and turned to the purposes of the Institution which he served. Rumford was no longer its moving and controlling spirit; his duty to the Elector of Bavaria, and his ill-starred devotion to Madame Lavoisier, had gradually drawn him away from London, and in 1803 he ceased to take any active part in the fortunes of his offspring. Shortly afterwards Sir Joseph Banks also withdrew. In a letter written April, 1804, he tells Rumford that his continued absence from England is a great detriment to the Institution:

It is now entirely in the hands of the profane. I have declared my dissatisfaction at the mode in which it is carried on, and my resolution not to attend in future. Had my health and spirits not failed me, I could have kept matters in their proper level, but sick, alone, and unsupported, I have given up what cannot now easily be recovered.

Sir John Hippesley, who became treasurer, strove to make the Institution above all things fashionable. He had a project for placing private boxes in the theatre, and was concerned about its want of a proper coat-of-arms. Mr Bernard still continued to hope that Sydney Smith's lectures on Moral Philosophy might somehow better the condition of the poor. They would, at least, said Homer, "make the real blue-stockings a little more disagreeable than ever, and sensible women a little more sensible." But the real directing power was Davy, and he gradually stamped upon the place the character it now possesses. How he felt his power and used it, may be gleaned from the following extract from a lecture in 1809, in reference to a fund which had been raised to supply him with a great voltaic battery:

In a great country like this, it was to be expected that a fund could not long be wanting for pursuing or perfecting any great Scientific object. But the promptitude with which the subscription filled was so great, as to leave no opportunity to many zealous patrons of science for showing their liberality. The munificence of a few individuals has afforded means more ample and magnificent than those furnished by the Government of a rival nation; and I believe we have preceded them in the application of the means. In this kind of emulation, our superiority, I trust, will never be lost; and I trust that the activity belonging to our sciences will always flow from the voluntary efforts of individuals, from whom the support will be an honour—to whom it will be honourable ...

Without facilities for pursuing his object, the greatest genius in experimental research may live and die useless and unknown. Talents of this kind cannot, like talents for literature and the fine arts, call forth attention and respect. They can neither give popularity to the names of patrons, nor ornament their houses. They are limited in their effects, which are directed towards the immutable interests of society. They cannot be made subservient to fashion or caprice; they must forever be attached to truth, and belong to nature. If we merely consider *instruction* in physical science, this even requires an expensive apparatus to be efficient; for without proper ocular demonstrations, all lectures must be unavailing,—things rather than words should be made the objects of study. A certain knowledge of the beings and substances surrounding us must be felt as a want by every cultivated mind. It is a want which no activity of thought, no books, no course of reading or conversation can supply. That a spirit for promoting experimental science is not wanting in the country, is proved by the statement which I have just made, by the foundation in which I have the honour of addressing you, and by the number of institutions rising in different parts of the metropolis and in the provinces. But it is clear that this laudable spirit may produce little effect from want of just direction. To divide and to separate the sources of scientific interest, is to destroy all their just effect. To attempt, with insufficient means, to support philosophy, is merely to humiliate her and render her an object of derision. Those who establish foundations for teaching the sciences ought, at least, to understand their dignity. To connect pecuniary speculation, or commercial advantages, with schemes for promoting the progress of knowledge, is to take crops without employing manure; is to create sterility, and to destroy improvement. A scientific institution ought no more to be made an object of profit than an hospitable, or a charitable establishment. Intellectual wants are at least as worthy of support as corporeal wants, and they ought to be provided for with the same feeling of nobleness and liberality. The language expected by the members of a scientific body from the directors ought not to be, 'We have increased your property, we have raised the value of your shares.' It ought rather to be, 'We have endeavoured to apply your funds to useful purposes, to promote the diffusion of science, to encourage discovery, and to exalt the scientific glory of your country.'

What this institution has done, it would ill become a person in my place to detail; but that it has tended to the progress of knowledge and invention, will not, I believe, be questioned. Compare the expenditure with the advantages. It would not support the least of your public amusements; and the income of an establish which, in its effects, may be said to be national, is derived from annual subscriptions scarcely greater than those which a learned professor of Edinburgh obtains from a single class ...

The progression of physical science is much more connected with your prosperity than is usually imagined. You owe to experimental philosophy some of the most important and peculiar of your advantages. It is not by foreign conquests chiefly that you are become great, but by a conquest of nature in your own country. It is not so much by colonization that you have attained your pre-eminence or wealth, but by the cultivation of the riches of your own soil. Why, at this moment, are you able to supply the world with a thousand articles of iron and steel necessary for the purposes of life? It is by arts derived from chemistry and mechanics, and founded purely upon experiments. Why is the steam engine now carrying on operations which formerly employed, in painful and humiliating labour, thousands of our robust peasantry, who are now more nobly or more usefully serving their country either with the sword or with the plough? It was in consequence of experiments upon the nature of beat and pure physical investigations.

In every part of the world manufactures made from the mere clay and pebbles of your soil may be found; and to what is this owing? To chemical arts and experiments. You have excelled all other people in the products of industry. But why? Because you have assisted industry by science. Do not regard as indifferent what is your true and greatest glory. Except in these respects, and in the light of a pure system of faith, in what are you superior to Athens or to Rome? Do you carry away from them the palm in literature and the fine arts? Do you not rather glory, and justly too, in being, in these respects, their imitators? Is it not demonstrated by the nature of your system of public education, and by your popular amusements? In what, then, are you their superiors? In every thing connected with physical science ; with the experimental arts. These are your characteristics. Do not neglect them. You have a Newton, who is the glory, not only of your own country, but of the human race. You have a Bacon, whose precepts may still be attended to with advantage. Shall Englishmen slumber in that path which these great men have opened, and be overtaken by their neighbours? Say, rather, that all assistance shall be given to their efforts that they shall be attended to, encouraged, and supported.

On a subsequent occasion, when the subjugation of Europe was threatened by the restless military spirit of France, he thus dilated upon the influence of experimental philosophy in strengthening the desire for rational freedom:

The scientific glory of a country may be considered, in some measure, as an indication of its innate strength. The exaltation of reason must necessarily be connected with the exaltation of the other noble faculties of the mind; and there is one spirit of enterprise, vigour, and conquest, in science, arts, and arms.

Science for its progression requires patronage,—but it must be a patronage bestowed, a patronage received, with dignity. It must be preserved independent. It can bear no fetters, not even fetters of gold, and least of all those fetters in which ignorance or selfishness may attempt to shackle it.

And there is no country which ought so much to glory in its progress, which is so much interested in its success, as this happy island. Science has been a prime cause of creating for us the inexhaustible wealth of manufactures, and it is by science that it must be preserved and extended. We are interested as a commercial people,—we are interested as a free people. The age of glory of a nation is likewise the age of its security. The same dignified feeling, which urges men to endeavour to gain a dominion over nature, will preserve them from the humiliation of slavery. Natural, and moral, and religious knowledge, are of one family; and happy is that country, and great its strength, where they dwell together in union.

It was, of course, to be expected that amidst the general chorus of approval some discordant notes should be heard. People who preferred the severe and formal manner of his colleague, Dr Young, who, in spite of his profound knowledge, could never keep an audience together, said that Davy's style was too florid and imaginative; that his imagery was inappropriate, and his conceits violent; that he was affected and swayed by a mawkish sensibility. Dr Paris would have us believe there was some show of justice in this accusation, but he thinks that "the style which cannot be tolerated in a philosophical essay may under peculiar circumstances be not only admissible but even expedient in a popular lecture." The "peculiar circumstance" in Davy's case was, in Dr Paris's opinion, the Royal Institution audience.

Let us consider for a moment," he says, "the class of persons to whom Davy addressed himself. Were they students prepared to toil with systematic precision, in order to obtain knowledge as a matter of necessity?—No—they were composed of the gay and the idle, who could only be tempted to admit instruction by the prospect of receiving pleasure,—they were children, who could only be induced to swallow the salutary draught by the honey around the rim of the cup.

That Davy himself was not wholly unconscious of this fact may be gathered from a letter which he wrote to Mr Davies Gilbert at about this time. He says:

My labours in the Theatre of the Royal Institution have been more successful than I could have hoped from the nature of them. In lectures, the effect produced upon the mind is generally transitory; for the most part, they amuse rather than instruct, and stimulate to enquiry rather than give information. My audience has often amounted to four and five hundred, and upwards; and amongst them some promise to become permanently attached to chemistry. This science is much the fashion of the day.

Whatever may be urged against Davy's style of lecturing, his purely scientific memoirs are unquestionably models of their kind. His language is so simple, and his mode of expression so uniformly clear, and so free from technicality, that even an ordinary reader can follow them with delight. In this respect he was consistently faithful to the direction he gives in his *Last Days*:

In detailing the results of experiments, and in giving them to the world, the chemical philosopher should adopt the simplest style and manner; he will avoid all ornaments, as something injurious to his subject, and should bear in mind the saying of the first King of Great Britain, respecting a sermon which was excellent in doctrine, but overcharged with poetical allusions and figurative language,— 'that the tropes and metaphors of the speaker were like the brilliant wild flowers in a field of corn, very pretty, but which did very much hurt the corn.'

Dr Paris's remarks concerning Davy's personal manner and his style of lecturing were warmly controverted at the time of their publication by several of Davy's friends. Dr John Davy's account is so clear and explicit, and so obviously based upon personal observation, for which he had ample opportunities, that, even after making every allowance for brotherly bias, we prefer to regard it as giving a more just impression of Davy's bearing in the lecture-theatre, and of the care and pains he took to ensure success.

"He was," says Dr Davy, "always in earnest; and when he amused most, amusement appeared most foreign to his object. His great and first object was to instruct, and, in conjunction with this, maintain the importance and dignity of science; indeed the latter, and the kindling a taste for scientific pursuits, might rather be considered his main object, and the conveying instruction a secondary one."

His lecture was almost invariably written expressly for the occasion, and usually on the day before he delivered it.

On this day he generally dined in his own room, and made a light meal on fish. He was always master of his subject; and composed with great rapidity, and with a security of his powers never failing him … It was almost an invariable rule with him, the evening before, to rehearse his lecture in the presence of his assistants, the preparations having been made and everything in readiness for the experiments ; and this he did, not only with a view to the success of the experiments, and the dexterity of his assistants, but also in regard to his own discourse, the effect of which, he knew, depended upon the manner in which it was delivered. He used, I remember, at this recital, to mark the words which required emphasis and study the effect of intonation; often repeating a passage two or three different times, to witness the difference of effect of variations in the voice. His manner was perfectly natural, animated and energetic, but not in the least theatrical. In speaking, he never seemed to consider himself as an object of attention; he spoke as if devoted to his subject, and as if his audience were equally devoted to it and their interest concentrated in it. The impressiveness of his oratory was one of its great charms … and his eloquence,—the declamation, as it might be called by some, in which he indulged on the beauty and order of Nature … was so well received because it was not affected; merely his own strong impressions and feelings embodied in words, and delivered with an earnestness which marked their sincerity.

It must, however, be admitted that this extraordinary success was not without its evil influence on Davy's moral qualities. Considering his age, and his temperament, his ambition and love of applause, he would have been something more than human if he could have remained wholly unaffected by the conditions in which he was placed. "The bloom of his simplicity was dulled by the breath of adulation." He assumed the garb and the airs of a man of fashion, and courted the society of the rich and the aristocratic. Time which would have been more profitably spent in the study, or in the society of his intellectual fellows, was frittered away in the frivolities of London society, or in the *salons*, or at the *soirées* of leaders of the "smart" people of the period. The peculiar circumstances of the Royal Institution, and the necessity for the continued adhesion to it of persons of rank and wealth, may to some extent have led him away from the quieter and serener joys of the philosophic life.

"In the morning," says Paris, "he was the sage interpreter of Nature's laws; in the evening, he sparkled in the galaxy of fashion; and not the least extraordinary point in the character of this great man, was the facility with which he could cast aside the cares of study, and enter into the trifling amusements of society.—'*Ne otium quidem otiosum*,' was the exclamation of Cicero; and it will generally apply to the leisure of men actively engaged in the pursuits of science; but Davy, in closing the door of his laboratory, opened the temple of pleasure … In ordinary cases, the genius of evening dissipation is an arrant Penelope; but Davy, on returning to his morning labours, never found that the thread had been unspun during the interruption."

The following letter from Coleridge will serve to show how this change was foreseen and deplored by his truest friends:

Nether Stowey, *Feby.* 17, 1803.

MY DEAR PURKIS, … I have been here nearly a fortnight; and in better health than usual. Tranquillity, warm rooms and a dear old friend, are specifics for my complaints. Poole is indeed a very, very good man. I like even his incorrigibility in small faults and deficiencies ; it looks like a wise determination of Nature to let well alone; and is a consequence, a necessary one perhaps, of his immutability in his important good qualities …

I rejoice in Davy's progress. There are three suns recorded in Scripture: Joshua's, that stood still; Hezekiah's, that went backward; and David's that went forth, and hastened on his course, like a bridegroom from his chamber. May our friend's prove the latter. It is a melancholy thing to see a man, like the Sun in the close of the Lapland summer, meridional in his horizon; or like wheat in a rainy season, that shoots up well in the stalk, but does not *kern*. As I have hoped, and do hope, more proudly of Davy than of any other man; and as he has been endeared to me more than any other man, by the being a Thing of Hope to me (more, far more than myself to my own self in my most genial moments,)—so of course my disappointment would be proportionally severe. It were falsehood, if I said that I think his present situation most calculated, of all others, to foster either his genius, or the clearness and incorruptness of his opinions and moral feelings. I see two Serpents at the cradle of his genius: Dissipation with a perpetual increase of acquaintances, and the constant presence of Inferiors and Devotees, with that too great facility of attaining admiration which degrades Ambition into Vanity—but the Hercules will strangle both the reptile monsters. I have thought it possible to exert talents with perseverance, and to attain true greatness wholly pure, even from the impulses; but on this subject Davy and I always differed …

Yours sincerely
S.T. COLERIDGE.

It would seem that Coleridge's doubts and fears were shared also by his host, and were communicated by him to the object of them. This, at least, may be inferred from the following extract from a letter from Davy to Poole:

London, *May* 1, 1803.
MY DEAR POOLE … Be not alarmed, my dear friend, as to the effect of worldly society on my mind. The age of danger has passed away. There are in the intellectual being of all men, permanent elements, certain habits and passions that cannot change. I am a lover of Nature, with an ungratified imagination. I shall continue to search for untasted charms—for hidden beauties.

My *real*, my *waking* existence is amongst the objects of scientific research : common amusements and enjoyments are necessary to me only as dreams, to interrupt the flow of thoughts too nearly analogous to enlighten and to vivify.

Coleridge has left London for Keswick; during his stay in town, I saw him seldomer than usual; when I did see him, it was generally in the midst of large companies, where he is the image of power and activity. His eloquence is unimpaired; perhaps it is softer and stronger. His will is probably less than ever commensurate with his ability. Brilliant images of greatness float upon his mind like the images of the morning clouds upon the waters, their forms are changed by the motion of the waves, they are agitated by every breeze, and modified by every sunbeam. He talked in the course of one hour, of beginning three works, and he recited the poem of Christabel unfinished, and as I had before heard it. What talent does he not waste in forming visions, sublime, but unconnected with the real world I have looked to his efforts, as to the efforts of a creating being; but as yet, he has not even laid the foundation for the new world of intellectual form …

Your affectionate friend
HUMPHRY DAVY.

Space will not permit of any more detailed account of Davy's career as a lecturer at the Royal Institution. During the twelve years he occupied its Chair of Chemistry he held undisputed sway as the greatest living expositor of chemical doctrine, and session after session saw the theatre crowded with eager and expectant audiences.

This continued and increasing success was due not merely to his art and skill as a speaker, but to the remarkable and astonishing character of what he had to tell—of work which made the laboratory of the Royal Institution even more famous than its lecture-rooms.

---

1. Young's connection with the Royal Institution was comparatively brief. On July 4th, 1803, it was resolved "That Dr Young be paid the balance of two years' complete salary, and that his engagement with the Institution terminate from this time."

V

THE CHEMICAL LABORATORY OF THE
ROYAL INSTITUTION

THE CHEMICAL LABORATORY OF THE Royal Institution, as the scene of
Davy's greatest discoveries—-discoveries which mark epochs in the
development of natural knowledge— will for ever be hallowed ground to
the philosopher. The votaries of Hermes have raised far more stately temples;
today they follow their pursuit in edifices which in architectural elegance and
in equipment are palaces compared with the subterranean structure which
lies behind the Corinthian façade in Albemarle Street. But to the chemist this
spot is what the Ka'ba at Mecca is to the follower of Mohammed, or what
Iona was to Dr Johnson; and, if we may venture to adapt the language of the
English moralist, that student has little to be envied whose enthusiasm would
not grow warmer or whose devotion would not gain force within the place
made sacred by the genius and labours of Davy and Faraday.

The first year of the century is memorable for the invention of the voltaic
pile, and for the discovery, by Nicholson and Carlisle, on April 30th, 1800,
of the electrolytic decomposition of water. As Davy said, "the voltaic battery
was an alarm-bell to experimenters in every part of Europe; and it served no
less for demonstrating new properties in electricity, and for establishing the
laws of this science, than as an instrument of discovery in other branches of
knowledge; exhibiting relations between subjects before apparently without
connection, and serving as a bond of unity between chemical and physical
philosophy." The capital discovery of Volta was made known in England at
the earliest possible moment through the mediation of Sir Joseph Banks, and
the study of voltaic electricity, its effects and applications, was immediately
afterwards entered upon by many English men of science with great zeal and

ardour. Davy at this time had just completed his work on Nitrous Oxide; and, powerfully impressed with the significance of Nicholson and Carlisle's observation, he at once turned his attention to the subject, and even before leaving Bristol he had sent a number of short papers on what was then usually termed the galvanic electricity to Nicholson's Journal. He showed that oxygen and hydrogen were evolved from separate portions of water, though vegetable and even animal substances intervened; and conceiving that all decomposition might be polar, he "electrised" different compounds at the different extremities, and found that sulphur and metallic substances appeared at the negative pole, and oxygen and nitrogen at the positive pole, though the bodies furnishing them were separate from each other. The papers, however, are mainly remarkable for the fact that they served to establish the intimate connection between the electrical effects and the chemical changes going on in the pile, and for the conclusion drawn concerning their mutual dependence. Within a few days after his removal to the Royal Institution he resumed his inquiries, publishing his results in a series of notices in the short-lived Journal of the Royal Institution.

In 1801 he sent his first communication to the Royal Society, on "An Account of some Galvanic Combinations, formed by the Arrangement of single metallic Plates and Fluids, analogous to the new Galvanic Apparatus of Mr Volta."

But at this period, and for some time afterwards, Davy was not altogether free to develop his own ideas, as the work of the laboratory was controlled by a committee which met, from time to time, to deliberate and settle upon the researches which were to be undertaken by their Professor. As we have seen, he was requested, in the first place, to turn his attention to tanning, and to investigate the astringent principles employed in the manufacture of leather. Afterwards, when the Managers determined to form a mineralogical collection, and to institute an assay office for the improvement of mineralogy and metallurgy, he was ordered to make analyses of rocks and minerals. And lastly, in consequence of an arrangement between the Managers and the Board of Agriculture, effected by Arthur Young, he was required to take up the subject of Agricultural Chemistry. To a man of Thomas Young's temperament the fussy activity of committees, directed by such people as Bernard and Hippesley, would have been resented as an irksome, if not intolerable, interference; but Davy invariably acted as if he considered that their decisions promoted the true interests of the Institution, and entered with ardour into each new scheme. There was no irksomeness to him in being called upon to change the current of his ideas, for he delighted in the opportunity of exhibiting his versatility; and, confident in his powers, he had the ambition to touch everything in turn, and to adorn it. That he should have succeeded so well under such conditions is perhaps the strongest evidence that could be adduced of the strength and elasticity of his eager, active mind, and of his astonishing power of rapid, well-directed work.

We have already dealt with his researches in connection with tanning. The efforts of the Managers towards the improvement of mineralogy and metallurgy, in spite of the generous assistance of Mr Greville, Sir J. St Aubin, and Sir A. Hume, and the "activity and intelligence of Mr Davy," proved abortive.

One outcome of Davy's association with the matter may be seen in his paper, published by the Royal Society in 1805, on "An Account of some analytical Experiments on a mineral Production from Devonshire, consisting principally of Alumine and Water." The mineral referred to was discovered by Dr Wavel in an argillaceous slate near Barnstaple, and hence was termed *wavellite*. Davy failed to recognise its true nature, which was first correctly ascertained by Berzelius. A few weeks later, he sent to the Royal Society a second paper "On a Method of Analyzing Stones containing fixed Alkali, by Means of the Boracic Acid." The method, however, is of comparatively limited application, and is seldom, if ever, now used in analysis. Determinative chemistry was never one of Davy's strong points, and few of his analytical processes are now employed. Patient manipulation, and minute and sustained attention to detail, were altogether foreign to his disposition and habits, although he had the highest appreciation of these qualities in men like Cavendish and Wollaston.

The lectures on agriculture however, were a great success, and brought increased fame and no small profit to the lecturer. His association with the Board of Agriculture developed into a permanent appointment; for ten successive years he continued to lecture on the subject before its members, and in 1813 he put together the results of his labours in his well-known *Elements of Agricultural Chemistry*. In simplicity and absence of ornament the style of these lectures is in marked contrast to that which he usually employed at the Royal Institution. Dealing with men to whom the matter was of paramount importance, he had no need to stimulate their interest by the arts he employed in the theatre in Albemarle Street. The very nature of the subject, perhaps, served to remind him that tropes and metaphors were here as much out of place as "the brilliant wild flowers in the field of corn—very pretty, but which did very much hurt the corn."

It would be impossible in the space at our disposal to attempt to give a minute analysis of Davy's work in connection with agriculture. Its interest now is, for the most part, historical; what is of permanent importance in the way of fact has long since been woven into the common web of knowledge. Its greatest value was not in the novelty or the abundance of its facts, but rather as a closely-reasoned exposition of the relation of agriculture to science, and of the necessity for applying the principles and methods of science to the art. The philosophic breadth of his views, supported, on occasion, by apt example and striking analogy, might be illustrated by many extracts. This, for example, is how he speaks of the value of the scientific method, and of chemistry, to husbandry:

Nothing is more wanting in agriculture than experiments, in which all the circumstances are minutely and scientifically detailed. This art will advance with rapidity in proportion as it becomes exact in its methods. As in physical researches all the causes should be considered; a difference in the results may be produced, even by the fall of a half an inch of rain more or less in the course of a season, or a few degrees of temperature, or even by a slight difference in the subsoil, or in the inclination of the land.

Information collected, after views of distinct inquiry, would necessarily be more accurate, and more capable of being connected with the general principles of science; and a few histories of the results of truly philosophical experiments in agricultural chemistry would be of more value in enlightening and benefiting the far than the greatest possible accumulation of imperfect trials conducted merely in the empirical spirit. It is no unusual occurrence, for persons who argue in favour of practice and experience, to condemn generally all attempts to improve agriculture by philosophical inquiries and chemical methods. That much vague speculation may be found in the works of those who have lightly taken up agricultural chemistry, it is impossible to deny. It is not uncommon to find a number of changes rung upon a string of technical terms, such as oxygen, hydrogen, carbon, and azote, as if the science depended upon words rather than upon things. But this is, in fact, an argument for the necessity of the establishment of just principles of chemistry on the subject. Whoever reasons upon agriculture, is obliged to recur to this science. He feels that it is scarcely possible to advance a step without it; and if he is satisfied with insufficient views, it is not because he prefers them to accurate knowledge, but, generally, because they are more current … It has been said, and undoubtedly with great truth, that a philosophical chemist would most probably make a very unprofitable business of farming; and this certainly would be the case, if he were a mere philosophical chemist; and unless he had served his apprenticeship to the practice of the art, as well as to the theory. But there is reason to believe that he would be a more successful agriculturist than a person equally uninitiated in farming, but ignorant of chemistry altogether; his science, as far as it went, would be useful to him. But chemistry is not the only kind of knowledge required it forums a part of the philosophical basis of agriculture; but it is an important part, and whenever applied in a proper manner must produce advantages.

How highly these lectures were appreciated will be evident from the terms in which they were referred to by Sir John Sinclair in his address of 1806 to the Board. He says:

> In the year 1802, when my Lord Carrington was in the chair, the Board resolved to direct the attention of a celebrated lecturer, Mr Davy, to agricultural subjects; and in the following year, during the presidency of Lord Sheffield, he first delivered to the members of this Institution, a course of lectures on the Chemistry of Agriculture. The plan has succeeded to the extent which might have been expected from the abilities of the gentleman engaged to carry it into effect. The lectures have hitherto been exclusively addressed to the members of the Board; but to such a degree of perfection have they arrived, that it is well worthy of consideration, whether they ought not to be given to a larger audience.

The "degree of perfection" was in no small degree due to the amount of experimental and observational work which Davy introduced into his lectures. Mr Bernard allotted him a considerable piece of ground on his property at Roehampton for experimental purposes, and the Duke of Bedford carried out trials for him at Woburn. He studied from time to time all the operations of practical farming, examined a great variety of soils, and investigated the nature and action of manures. He was thus brought into contact with some of the largest landowners and agriculturists of his time, and was an honoured guest in the houses of men like Lord Sheffield, Lord Thanet, Mr Coke of Holkham, and others.[1] In the practical interest he thus displayed in the most useful of all the arts he sought to emulate the example of his illustrious prototype Lavoisier, and his work constitutes the foundation of every treatise on the subject since the appearance, in 1840, of Liebig's well-known book.

Professor Warington, than whom no one is more fitted to express an opinion, has favoured me with the following critical estimate of the value of Davy's work:

> The lectures profess to be exhaustive and thus present all that Davy had been able to collect on the subject of the relations of chemistry to agriculture dur-

ing a period of at least 10 years. He appears to have made a careful study of the problems of agriculture for many years, and to be acquainted with English practice, and English experiments. There is but little reference to foreign practice, or foreign opinion, save where the work done has been purely chemical, as *e.g.* that of Gay Lussac, or Vauquelin. He approaches his subject in a thoroughly scientific manner, taking an independent view of each question, bringing all the knowledge at his disposal to bear upon it, and not hesitating to come to conclusions different from those usually received. The *great step* taken in these lectures is the assertion that Agriculture must look to Natural Science, and especially to Chemistry, for the explanation of its problems and the improvement of its practice. Davy seems to have been the first, at least in this country, who boldly claimed for 'Agricultural Chemistry' the position of a distinct branch of science. He was probably the earliest example of a first-class chemist, who seriously and continuously devoted his best attention to the subject of agriculture.

The lectures, looked at from a modern standpoint, are of unequal value. The method of food-analysis is very poor, and it is somewhat surprising that the accurate mode of determining nitrogen employed by Gay Lussac is not made use of in Davy's analyses. Nevertheless he manages to ascertain that spring sown wheat is richer in gluten than autumn sown, and the wheat of hot countries richer than the wheat of temperate regions, statements which are quite correct.

Lecture VI is decidedly poor. Davy believes that plants feed on carbonaceous matter by their roots, and this mistaken theory leads him to assign an undue value to organic substances as manures, It seems curious nowadays to find the whole subject of manures treated with hardly any reference to their contents in nitrogen, phosphoric acid, or potash.

Lecture IV is one of his best lectures, full of keen observation and suggestive experiment.

The references to his own agricultural experiments are very numerous; he seems to have made experiments on every subject of inquiry that came before him. There is however no attempt at an extended and thorough investigation of any subject, and for want of this the truth is sometimes missed. Thus in his trials of various ammonium salts as manures he finds the carbonate to be effective, the chloride to be of little value, and the sulphate of no good at all, whereas the last-named salt is now generally chosen as a manure.

There are some paragraphs that read like the inspirations of genius, though it is now of course difficult to tell to what extent his statements and opinions were warranted by the facts then known. He gives a wonderfully correct idea of the action of peas or beans in rotation, even including the statement that they obtain their nitrogen from the atmosphere.

Although his time and energy were necessarily largely absorbed by the demands of the Managers, Davy never lost sight of the subject of voltaic electricity, and at intervals he was able to resume his inquiries upon it. What specially impressed him was the power of the voltaic pile as an analytic agent; and his laboratory journals, preserved at the Royal Institution, record the results of numerous trials on the behaviour of compound substances under its influence. In spite of innumerable distractions and constant interruptions, due mainly to the precarious position of the Institution, Davy gradually succeeded in unravelling the fundamental laws of electro-chemistry, and in thus importing a new order of conceptions, altogether unlooked for and undreamt of, into science. This really constitutes his greatest claim as a philosopher to our admiration and gratitude. The isolation of the metals of the alkalis, and the proof of the compound nature of the alkaline earths, were unquestionably achievements of the highest brilliancy, and as such appeal strongly to the popular imagination. But they were only the necessary and consequential links in a chain of discovery which, had Davy neglected to make them, would have been immediately forged by others. It is significant that almost immediately after the capital discovery of Nicholson and Carlisle, Dr Henry of Manchester, the well-known friend and collaborator of Dalton, should have made the attempt to separate the presumed metallic principle of potash by the agency of voltaic electricity.

Davy communicated the results of his inquiries made prior to the summer of 1806 in a paper to the Royal Society, which was made the Bakerian lecture of the year.[2] It is entitled "On some chemical Agencies of Electricity," and is divided in nine sections and an introduction. In the first section, "On the Changes produced by Electricity in Water," he set at rest the disputed question as to the origin of the acid and alkaline matter which had been observed to form during the electrolysis of this liquid. By some these substances were supposed to be *generated* from pure water by the action of electricity; and M. Brugnatelli had even attempted to prove the existence of a body *sui generis* which he termed the *electric acid*. By a series of Convincing experiments Davy showed that the substances were due to the presence of saline matter in the water, derived either from faulty purification, or from the solvent action of the water on the vessels, etc., with which it was in contact. Cruickshank had found that in some cases the acid was nitric acid and the alkali ammonia: these substances were shown by Davy to be due to the presence of dissolved air. When pure water, contained in vessels on which it exerted no solvent action, was "electrised" *in vacuo*, not a trace of either acid or alkali was produced.

In the second section, "On the Agencies of Electricity in the Decomposition of various Compounds," he begins by pointing out that in all the experiments recorded in the preceding section—that is, in all changes in which acid and alkaline matter had been present—the acid matter collected in the water round the positive pole, and the alkaline matter round the negative pole. This he shows to be true even of such sparingly soluble substances as gypsum, the sulphates of strontium and barium, and fluorspar. By connecting together cups or vessels made of the substances under investigation by a thread of well-washed asbestos, as suggested by Wollaston, he found that in all cases the acid element collected round the positive, and the earthy base round the negative pole. Basalt from Antrim, a zeolite from the Giant's Causeway, vitreous lava from Etna, and even glass, in like manner yielded alkaline matter to water when subjected to the action of voltaic electricity. Soluble salts, such as the sulphates of sodium, potassium, and ammonium, the nitrates of potassium and barium, the succinate, oxalate and benzoate of ammonium, were similarly decomposed: the acids in a certain time collected in the tube containing the positive wire, and the alkalis and earths in that containing the negative wire. When metallic solutions, such as those of iron, zinc, and tin were employed, metallic crystals or depositions were formed on the negative wire, and oxide was, likewise deposited round it; and a great excess of acid was soon found in the opposite cup.

In the next section, "On the Transfer of Certain of the Constituent Parts of Bodies by the Action of Electricity," he points out that the observations of Gautherot and of Hisinger and Berzelius rendered it probable that the saline elements evolved in decompositions by electricity were capable of being transferred from one electrified surface to another, according to their usual order of arrangement, but that exact observations on this point were wanting. He connected a cup of gypsum with one of agate by means of asbestos, and filling each with purified water, he inserted the negative wire of the battery in the agate cup, and the positive wire in that of the sulphate of lime. In about four hours he found a strong solution of lime in the agate cup, and sulphuric acid in that of gypsum. By reversing the order, and carrying on the process for a similar length of time, the sulphuric acid appeared in the agate cup, and the solution of lime on the opposite side. Many trials were made with other saline substances with analogous results.

The time required for these transmissions (the quantity and intensity of the electricity, and other circumstances remaining the same) seemed to be related to the length of the intermediate column of water.

To ascertain whether the contact of the saline solution with a metallic surface was necessary for the decomposition and transference, he introduced purified water into two glass tubes; a vessel containing solution of potassium chloride was connected with each of the tubes by means of asbestos; on introducing the wires into the tubes alkaline matter soon appeared in one tube, and acid matter in the other; and in the course of a few hours moderately strong solutions of potash and of hydrochloric acid were formed.

Two tubes, one containing distilled water, the other a solution of potassium sulphate, were each connected by asbestos threads with a vessel containing a dilute solution of litmus; the saline matter was negatively electrified and as it was natural to suppose that the sulphuric acid in passing through the water to the positive side would redden the litmus in its course, some slips of litmus paper were placed above and below the pieces of asbestos, directly in the circuit: it was found that the acid and alkali passed through the litmus solution without effecting any change in colour.

As acid and alkaline substances during the time of their electrical transfer passed through water containing vegetable colours without affecting them, or apparently combining with them, it immediately became an object of inquiry whether they would not likewise pass through chemical menstrua having stronger attractions for them; and it seemed reasonable to suppose that the same power which destroyed elective affinity in the vicinity of the metallic points would likewise destroy it, or suspend its operation, throughout the whole of the circuit.

To test this supposition, solution of potassium sulphate was placed in contact with the negative wire, and pure water in contact with the positive wire and a weak solution of ammonia was made the middle link of the conducting chain, so that no sulphuric acid could pass to the positive pole in the distilled water without passing through the solution of ammonia.

In less than five minutes it was found that acid was collecting round the positive pole, and in half an hour the water was sour to the taste, and gave a precipitate with barium nitrate. Hydrochloric acid from common salt, and nitric acid from nitre were transmitted through concentrated alkaline menstrua under similar circumstances. Strontia and baryta readily passed, like the other alkaline substances, through hydrochloric and nitric acids; and vice versa these acids passed with facility through aqueous solution of baryta and strontia; but it was impossible to pass sulphuric acid through baryta or strontia, or to pass baryta and strontia through sulphuric acid, as precipitates of insoluble barium and strontium sulphate were formed.

In the next section, "On Some General Observations on these Phenomena, and on the Mode of Decomposition and Transition," he summarises the foregoing results:

> It will be a general expression of the facts that have been detailed, relating to the changes and transitions by electricity, in common philosophical language, to say that hydrogen, the alkaline substances, the metals, and certain metallic oxides, are attracted by negatively electrified metallic surfaces, and repelled by positively electrified metallic surfaces; and contrariwise, that oxygen and acid substances are attracted by positively electrified metallic surfaces, and repelled by negatively electrified metallic surfaces and these attractive and repulsive forces are sufficiently energetic to destroy or suspend the usual operation of elective affinity.
>
> It is very natural to suppose, that the repellent and attractive energies are communicated from one *particle to another particle* of the same kind, so as to establish a conducting chain in the fluid; and that the locomotion takes place in consequence and that this is really the case seems to be shown by many facts. Thus, in all the instances in which I examined alkaline solutions through which acids had been transmitted, I always found acid in them whenever any acid matter remained at the original source …
>
> In the cases of the separation of the constituents of water, and of solutions of neutral salts forming the whole of the chain, there may possibly be a succession of decompositions, and re-compositions throughout the fluid. And this idea is strengthened by the experiments on the attempt to pass barytes through sulphuric acid, and muriatic acid through solution of sulphate of silver, in which as insoluble compounds are formed and carried out of the sphere of the electrical action, the power of transfer is destroyed.

In the next section, "On the General Principles of the Chemical Changes produced by Electricity," he points out that it had been already shown by Bennet that many bodies brought into contact and afterwards separated exhibited *opposite* states of electricity; and that this observation had been confirmed and extended by Volta, who had supposed that it also takes place with regard to metals and fluids. In his paper in the *Philosophical Transactions* of 1801, the first he sent to the Royal Society, Davy had shown that when alternations of single metallic plates and acid and alkaline solutions were employed in the construction of voltaic combinations, the alkaline solutions always received the electricity from the metal, and the acid always transmitted it to the metal.

In the simplest case of electrical action, the alkali which receives electricity from the metal would necessarily, on being separated from it, appear positive, whilst the acid under similar circumstances would be negative; and these bodies, having respectively with regard to the metals that which may be called a positive and a negative electrical energy, in their repellent and attractive functions seem to be governed by laws the same as the common laws of electrical attraction and repulsion.

The seventh section treats of The Relations between the Electrical Energies of Bodies and their Chemical Affinities":

As the chemical attraction between two bodies seems to be destroyed by giving one of them an electrical state different from that which it naturally possesses; that is, by bringing it artificially into a state similar to the other, so it may be increased by exalting its natural energy. Thus, whilst zinc, one of the most oxidable of the metals, is incapable of combining with oxygen when negatively electrified in the circuit, even by a feeble power; silver, one of the least oxidable, easily unites to it when positively electrified and the same thing might be said of other metals. Amongst the substances that combine chemically, all those, the electrical energies of which are well known, exhibit opposite states; thus copper and zinc, gold and quicksilver, sulphur and the metals, the acid and alkaline substances, afford opposite instances; and supposing perfect freedom of motion in their particles or elementary matter, they ought according to the principles laid down, to attract each other in consequence of their electrical powers. In the present state of our knowledge it would be useless to attempt to speculate on the remote cause of the electrical energy, or the reason why different bodies, after being brought into contact should be found differently electrified; its relation to chemical affinity is however, sufficiently evident. May it not be identical with it, and an essential property of matter?

How Davy sought to elaborate a theory of chemical affinity on these facts will be sufficiently obvious from the following extracts:

Supposing two bodies, the particles of which are in different electrical states, and those states sufficiently exalted to give them an attractive force superior to the power of aggregation, a combination would take place which would be more or less intense according as the energies were more or less perfectly balanced; and the change of properties would be correspondently proportional

When two bodies repellent of each other act upon the same body with different degrees of the same electrical attracting energy, the combination would be determined by the degree; and the substance possessing the weakest energy would be repelled and this principle would afford an expression of the causes of elective affinity and the decompositions produced in consequence.

Or where the bodies having different degrees of the same energy, with regard to the third body, had likewise different energies with regard to each other, there might be such a balance of attractive and repellent powers as to produce a triple compound; and by the extension of this reasoning, complicated chemical union may be easily explained.

As the combined effect of many particles possessing a feeble electrical energy may be conceived equal or even superior to the effect of a few particles possessing a strong electrical energy, the same principle may explain the influence of mass action, as elucidated by Berthollet.

He conceives also that it may be possible to obtain a *measure* of chemical affinity founded upon the energy of the voltaic apparatus required to destroy the chemical equilibrium. He points out that, as light and heat are the common consequences of the restoration of the equilibrium between bodies in a high state of opposite electricities, so it is perhaps an additional circumstance in favour of his theory to state that heat and light are likewise the result of all intense chemical action. And as in certain forms of the voltaic battery when large quantities of electricity of low intensity act, heat is produced without light; so in slow combinations there is an increase of temperature without luminous appearance. The effect of heat in producing combination may, he assumes, be also explained according to these ideas. It not only gives more freedom of motion to the particles, but in a number of cases—*e.g.* tourmaline, sulphur, etc.—it seems to exalt the electrical energies of bodies.

In the eighth section he seeks to apply these principles to the mode of action of the voltaic pile, and to explain the nature of the changes which occur between the plates and the exciting fluid, and he points out that the theory in some measure reconciles the hypothetical principles of the action of the pile adopted by its inventor with the opinions concerning the chemical origin of galvanism held by the majority of British men of science at that period. At the same time, Davy argues that the facts are in contradiction to the assumption that chemical changes are the *primary* causes of the phenomena of galvanism. Moreover, in mere cases of chemical change—as in iron burning in oxygen, the deflagration of nitre with charcoal, the combination of potash with sulphuric acid, the amalgamation of zinc,—electricity is never exhibited.

In the concluding section he trusts that many applications of the general facts and principles thus indicated to the processes of chemistry, both in art and in nature, may suggest themselves to the philosophical inquirer. It is not improbable, he thinks, that the electric decomposition of the neutral salts in different cases may admit of economical uses. He is induced to hope that the new mode of analysis may lead to the discovery of the *true* elements of bodies:

> For if chemical union be of the nature which I have ventured to suppose, however strong the natural electrical energies of the elements of bodies may be, yet there is every probability of a limit to their strength: whereas the powers of our artificial instruments seem capable of indefinite increase.

Phenomena similar to those occurring in the voltaic cell must be produced in various parts of the interior strata of our globe, and it is very probable that many mineral formations have been materially influenced, or even occasioned, by such action. The electrical power of transference may serve to explain some of the principal and most mysterious facts in geology.

> Natural electricity has hitherto been little investigated, except in the case of its evident and powerful concentration in the atmosphere. Its slow and silent operations in every part of the surface will probably be found more immediately and importantly connected with the order and economy of nature; and investigations on this subject can hardly fail to enlighten our philosophical systems of the earth, and may possibly place new powers within our reach.

The publication of this paper exercised a profound sensation, both at home and abroad. Berzelius, years afterwards, spoke of it as one of the most remarkable memoirs that had ever enriched the theory of chemistry—and the praise is the more significant when it is remembered that Davy had thereby seemed to have taken possession of a field of inquiry which the Swedish chemist, who was only a year younger than Davy, had been among the first to enter. Still more significant was the action of the French Institute. Bonaparte, when First Consul, had announced to the Institute his intention of founding a medal "for the best experiment which should be made in the course of each year on the galvanic fluid," and had further expressed his desire to give the sum of sixty thousand francs *"à celui qui, par ses expériences et ses découvertes fera faire l'électricité et au galvanisme un pas comparable à celui qu'ont fait faire à ces sciences Franklin et Volta."* A committee of the Institute, consisting of Laplace, Halle, Coulomb, Hauy and Biot, was appointed to consider the best means of accomplishing the wishes of the First Consul, and twelve months after the publication of the Bakerian lecture they awarded its author the medal. Whether the Institute had the means of awarding the sixty thousand francs as well is more than doubtful, for it does not appear that the sum named by Bonaparte ever went beyond the promise of it. All that the Institute got for themselves was, as Maria Edgeworth said, "a rating all round in imperial Billingsgate." The two countries at this period were at war, and the feeling of animosity was most bitter. Of course, there were persons who said that patriotism should forbid the acceptance of the award. Davy's own view was more sensible and politic. "Some people," he said to his friend Poole, "say I ought not to accept this prize; and there have been foolish paragraphs in the papers to that effect; but if the two countries or governments are at war, the men of science are not. That would, indeed, be a civil war of the worst description: we should rather, through the instrumentality of men of science, soften the asperities of national hostility."

1. In the print of the "Woburn Sheep-Shearing," Davy is represented as one of a group comprising Mr Coke, Sir Joseph Banks, Sir John Sinclair, and Mr Arthur Young.

2. This lecture, which is one of the events of each session of the Royal Society, owes its origin to Mr Henry Baker, FRS, a learned antiquary and naturalist, who, by his will of July, 1763, bequeathed the sum of £100 to the Society, the interest of which was to be applied "for an oration or discourse to be spoken or read yearly by some one of the Fellows of that Society, on such part of Natural history or Experimental Philosophy, at such time, and in such manner, as the President and Council of the said Society for the time being, shall please to order and appoint." Baker died in 1774, and the bequest came into operation during the presidency of Sir John Pringle; and Peter Woulfe—one of the last of the English alchemists—was appointed to deliver the lecture, which he did for three successive years.

VI

THE ISOLATION OF THE METALS
OF THE ALKALIS

However devoted Davy might be to scientific investigation, he was no less mindful of the sacred claims of the long vacation. In the summer of 1805 he went to the Lake Country, where he met Scott in company with Wordsworth; and the occasion on which the party "climbed the dark brow of the mighty Helvellyn," and which gave rise to Scott's well-known poem, is thus referred to by Lockhart:

> This day they were accompanied by an illustrious philosopher [Davy] who was also a true poet—and might have been one of the greatest of poets had he chosen; and I have heard Mr Wordsworth say, that it would be difficult to express the feelings with which he, who so often had climbed Helvellyn alone, found himself standing on its summit with two such men as Scott and Davy.

But the greater part of this summer he spent in the north of Ireland, examining the extraordinary geological features of that district. Lady Brownrigg, the sister of the Bishop of Raphoe, has given a spirited little account of her impressions of his appearance and manner at that period. She was, she says, very young at the time.

> We had been invited (I say *we*, for I was then with the Bishop of Raphoe) by Dr Richardson to go to his cottage at Portrush, 'to meet the famous Mr Davy.' We arrived a short time before dinner. In passing through a room we saw a youth, as he appeared, who had come in from fishing, and who, with a little notebook, was seated in a window-seat, having left a bag, rod &c., on the ground. He

was very intent upon this little book, and we passed through unnoticed. We shook hands with our host and hostess, and prepared for dinner. I went into the drawing-room, under some little awe of this great philosopher, annexing to such a character at least the idea of an elderly grave gentleman, not perhaps, with so large a wig as Dr Parr, or so sententious a manner as Dr Johnson,—but certainly I never calculated on being introduced to the identical youth, with a little brown head, like a boy, that we had seen with his book, and who, when I came into the drawing-room was in the most animated manner recounting an adventure on the Causeway which had entertained him and from his manner of telling it was causing loud laughing in the whole room.

Davy also spent much of the summer of 1806 in Ireland, and the journal which he kept during his tour contains many interesting notes of his impressions of the country and the people. In the course of his journey he visited Edgeworthstown—"the moral and intellectual paradise of the author of 'Castle Rackrent,' as he calls it. That gifted lady tells her cousin Sophy Buxton that as the result her head "was stuffed full of geological and chemical facts." "Mr Davy," she adds, "is wonderfully improved since you saw him at Bristol; he has an amazing fund of knowledge upon all subjects, and a great deal of genius."

There was much in Davy's own temperament to make him understand and appreciate the Irish character; himself a man of quick impulse and active sympathy, he was profoundly moved by the spectacle of Ireland's political degradation. In a letter to his friend Poole, written after his return to London, he says:

I long very much for the intercourse of a week with you: I have very much to say about Ireland. It is an island which might be made a new and a great country. It now boasts a fertile soil, an ingenious and robust peasantry, and a rich aristocracy; but the bane of the nation is the equality of poverty amongst the lower orders. All are slaves, without the probability of becoming free; they are in the state of equality which the *sans cullottes* wished for in France; and until emulation, and riches, and the love of clothes and neat houses are introduced among them, there will be no permanent improvement.

Changes in political institutions can, at first, do little towards serving them; it must be by altering their habits, by diffusing manufactories, by destroying *middlemen*, by dividing farms, and by promoting industry by making the pay proportional to the work: but I ought not to attempt to say anything on the subject when my limits are so narrow; I hope soon to converse with you about it.

With the exception of a rapid journey into Cornwall, for the sake of seeing his family, he spent the greater part of the summer and autumn of 1807 in town. He had been made Secretary of the Royal Society in succession to Gray, and was obliged to be in or near London in order to see the *Philosophical Transactions* through the press. From the Laboratory Journal it would appear that he was occupied at this time on a variety of disconnected investigations such as the nature of Antwerp Blue, and the effect of electricity on flame. In a letter to Davies Gilbert, dated September 12th, he states that he has been a good deal engaged in experiments on distillation for revenue purposes.

Towards the end of this month, or during the first week of October, he resumed his experiments with the voltaic battery, and he was led to study its action on the alkalis. There is some evidence that he had attacked the same question at Bristol. In a notebook of that period, under date August 6th, 1800, is the following sentence: "I cannot close this notice without feeling grateful to M. Volta, Mr Nicholson, and Mr Carlisle, whose experience has placed such a wonderful and important instrument of analysis in my power"—evidently a jotting to be used in one of the short communications to Nicholson's Journal. This is immediately followed by "Query: Would not potash, dissolved in spirits of wine, become a conductor?" And he then gives an account of some experiments on the action of voltaic electricity on aqueous solutions of ammonia, caustic potash, and hydrochloric acid, which apparently led to the same result as that already obtained by Nicholson and Carlisle in the case of water.

It is difficult to determine whether he had any precise idea, in again attacking the problem, or any expectation of a definite result. In one of his lectures at the Royal Institution on Electro-Chemical Science, delivered some time subsequently, he said he had a suspicion at that time that potash might turn out to be "phosphorus, or sulphur united to nitrogen":

> For as the volatile alkali was regarded as composed of an extremely light inflammable body—hydrogen—united to nitrogen, I conceived that *phosphorus* and *sulphur*, much denser bodies, might produce denser alkaline matter; and as there were no *known* combination of these with *nitrogen* it was probable that there might be unknown combinations.

Davy once said that "analogy was the fruitful parent of error"; and the whole history of science probably furnishes no more extraordinary instance of perverted analogy, or one more unexpected in its consequences. In another of his lectures he said of the alchemists that "even their *failures* developed some unsought-for object partaking of the marvellous"—and the statement in this case is even more true of himself. Each phase in the story of this discovery indeed partakes of the marvellous. Sometime during the first fortnight in October, 1807, he obtained his first decisive result; and on the 19th of November he delivered what is generally regarded as the most memorable of all his Bakerian lectures, "On some new Phenomena of chemical Changes produced by Electricity, particularly the Decomposition of the fixed Alkalies, and the Exhibition of the new substances which constitute their bases; and on the general Nature of alkaline Bodies." Few discoveries of like magnitude have been made and perfected in so short a time, and few memoirs have been more momentous in result than that which Davy put together in a few hours, and in which he announced his results to the world. The whole work was done under conditions of great mental excitement. His cousin Edmund Davy, who at the time acted as his assistant, relates that when he saw the minute globules of the quicksilver-like metal burst through the crust of potash and take fire, his joy knew no bounds; he actually danced about the room in ecstasy, and it was some time before he was sufficiently composed to continue his experiments. The rapidity with which he accumulated results after this first feeling of delirious delight had passed was extraordinary. Before the middle of November he had obtained most of the leading facts. In a letter dated November 13th he tells W. H. Pepys:

> I have decomposed and recomposed the fixed alkalies, and discovered their bases to be two new inflammable substances very like metals; but one of them lighter than ether, and infinitely combustible. So that there are two bodies decomposed, and two new elementary bodies found.

The stories told by Paris of his habits at this period, and of his various expedients to gain time—of his rushing off to dinner with persons of the highest rank with no fewer than five shirts on, and as many pairs of stockings, because in his haste he could not put on fresh linen without removing that which was underneath; of his continuing his chemical labours on his return to the laboratory until three or four in the morning; and of his then being up before the servants, are certainly much exaggerated, if not wholly apocryphal. He was, it is true, not very systematic in the disposal of his time, but he seldom entered the laboratory before ten or eleven in the morning, and rarely left it later than four, and he was scarcely ever known to visit it after he had dressed for dinner. Except when preparing a lecture, he seldom dined in his rooms at the Institution: his brother tells us that his invitations to dinner were so numerous that he was, or might have been, constantly engaged; and after dinner he was much in the habit of attending evening parties, and devoting the evening to amusement, "so that to the mere frequenters of such parties he must have appeared a votary of fashion rather than of science."

It was characteristic of him, that on the very eve of the announcement of the discovery which raised him to the summit of his scientific fame, he could unbend the strung bow and thus write to his youngest sister:

> MY DEAR SISTER ... I looked last week at the pattern of the gown that my sister put into my hands, and found it so worn and tattered that nothing can be made of it; I cannot therefore get your gowns made till you send me another. The best way will be to give me measure of the waist, shoulders, length &c., in this way, and there can then be no difficulties: thus waist, 15 inches, or whatever it may be; between shoulders: length from waist to skirt or train.
>
> I do not wish to send gowns you cannot wear, and in this way they can be well made. By a piece of tape you can easily measure and then try the length by a carpenter's rule, and give me the results for yourself, and for Kitty, and Grace, and I shall then be able to send your gowns a few days after I receive your letter.
>
> I shall write to my mother soon, about John. And now, my dear sister, having written you as stupid a letter as ever was written about gowns, I shall end with love to my mother, Kitty, Grace, and my aunts.
>
> Your affectionate brother
> H. DAVY.

The Bakerian lecture in which Davy announces the discovery of the compound nature of the fixed alkalis opens with a reference to the concluding remarks of his lecture of the previous year, "that the new methods of investigation promised to lead to a more intimate knowledge than had hitherto been obtained concerning the true elements of bodies. This conjecture, then sanctioned only by strong analogies, I am now happy to be able to support by some conclusive facts."

In the first attempts he made to decompose the fixed alkalis he acted upon concentrated aqueous solutions of potash and soda with the highest electrical power he could then command at the Royal Institution—*viz*, from voltaic batteries containing 24 plates of copper and zinc of 12 inches square, 100 plates of 6 inches, and 150 of 4 inches, charged with solutions of alum and nitric acid; but although there was high intensity of action nothing but hydrogen and oxygen was disengaged. He next tried potash in igneous fusion, and here the results were more encouraging: there were obvious and striking signs of decomposition; combustible matter was produced accompanied with flame and a most intense light. He had observed that although potash when dry is a non conductor, it readily conducts when it becomes damp by exposure to air, and in this state "fuses and decomposes by strong electrical powers."

> A small piece of pure potash, which had been exposed for a few seconds to the atmosphere, so as to give conducting power to the surface was placed upon an insulated disc of platina, connected with the negative side of the battery of the power of 250 of 6 and 4, in a state of intense activity[1] and a platina wire communicating with the positive side was brought in contact with the upper surface of the alkali …
>
> Under these circumstances a vivid action was soon observed to take place. The potash began to fuse at both its points of electrization. There was a violent effervescence at the upper surface; at the lower, or negative surface, there was no liberation of elastic fluid; but small globules having a high metallic lustre, and being precisely similar in visible characters to quicksilver appeared, some of which burnt with explosion and bright flame, as soon as they were formed, and others remained, and were merely tarnished, and finally covered by a white film which formed on their surfaces.

The platina, as such, was, he found, in no way connected with the result: a substance of the same kind was produced when copper, silver, gold, plumbago, or even charcoal was employed for completing the circuit.

> Soda when acted upon in the same manner as potash, exhibited an analogous result; but the decomposition demanded greater intensity of action in the batteries, or the alkali was required to be in much thinner and smaller pieces.
>
> The substance produced from potash remained fluid at the temperature of the atmosphere at the time of its production; that from soda, which was fluid in the degree of heat of the alkali during its formation, became solid on cooling, and appeared having the lustre of silver.

It would seem from his description of its properties that the potassium he obtained was most probably alloyed with sodium derived from impure potash. Potassium is solid up to 143° F.; but, as Davy subsequently found, an alloy of potassium and sodium is fluid at ordinary temperatures.

When the potassium was exposed to air its metallic lustre was immediately destroyed, and it was ultimately wholly reconverted into potash by absorption of oxygen and moisture.

With the substance from soda the appearance and effects were analogous.

When heated in oxygen to a sufficiently high temperature, both substances burnt with a brilliant white flame.

On account of their alterability on exposure to air, Davy had considerable difficulty in preserving and confining them so as to examine the properties of the new substances. As he says, like the *alcahests* imagined by the alchemists, they acted more or less upon almost every body to which they were exposed.

He eventually found that they might be preserved in naphtha.

The "basis" of potash at 50° F. was a soft and malleable solid with the lustre of polished silver.

> At about the freezing point of water it becomes harder and brittle, and when broken in fragments, exhibits a crystallized texture, which in the microscope seems composed of beautiful facets of a perfect whiteness and high metallic splendour.

It may be converted into vapour at a temperature approaching a red-heat, and may be distilled unchanged; it is a perfect conductor of electricity and an excellent conductor of heat. Its most marked difference from the common run of metals was its extraordinarily low specific gravity. Davy endeavoured to gain an approximation to its relative weight by comparing the weight of a globule with that of an equal-sized globule of mercury.

> Taking the mean of 4 experiments, conducted with great care, its specific gravity at 62° Fahrenheit, is to that of mercury as 10 to 223, which gives a proportion to that of water nearly as 6 to 10; so that it is the lightest fluid body known. In its solid form it is a little heavier.

Although no great stress can be laid on numbers so obtained, they serve to indicate that Davy had not yet obtained the pure metal. The real ratio of the specific gravities of potassium and mercury is as 10 to 154.

An account is then given of the behaviour of potassium towards oxygen, oxymuriatic acid gas [chlorine] hydrogen, water, alcohol, ether, the various mineral acids, phosphorus, sulphur, mercury, a number of metallic oxides, and the various forms of glass.

The "basis" of soda is described as a white opaque substance of the lustre and general appearance of silver. It is soft and malleable, and is a good conductor of heat and electricity. Its specific gravity was found by flotation in a mixture of oil of sassafras and naphtha to be 0.9348 (the true specific gravity of sodium is 0.974). It was found to fuse at about 180° F. (the real melting-point of sodium is 197.5°). Its action on a number of substances—oxygen, hydrogen, water, etc.—is then described, and its general behaviour contrasted with that of the "basis" of potash.

Davy then attempted, by synthetical experiments, to determine the amount of the "metallic bases" in potash and soda respectively, and, considering the extremely small quantities he had to operate upon, the results are fairly accurate.

He then enters upon some general observations on the relations of the "bases" of potash and soda to other bodies.

Should the bases of potash and soda be called metals? The greater number of philosophical persons to whom this question has been put, have answered in the affirmative. They agree with metals in opacity, lustre, malleability, conducting powers as to heat and electricity, and in their qualities of chemical combination.

Their low specific gravity does not appear a sufficient reason for making them a new class; for amongst the metals themselves there are remarkable differences in this respect ... and in the philosophical division of the classes of bodies, the analogy between the greater number of properties must always be the foundation of arrangement.

On this idea, in naming the bases of potash and soda, it will be proper to adopt the termination which, by common consent, has been applied to other newly discovered metals, and which, though originally Latin, is now naturalized in our language.

Potasium [sic] and sodium are the names by which I have ventured to call the new substances; and whatever changes of theory, with regard to the composition of bodies, may hereafter take place, these terms can scarcely express an error; for they may be considered as implying simply the metals produced from potash and soda. I have consulted with many of the most eminent scientific persons in this country, upon the methods of derivation, and the one I have adopted has been the one most generally approved. It is perhaps more significant than elegant. But it was not possible to found names upon specific properties not common to both,; and though a name for the basis of soda might have been borrowed from the Greek, yet an analogous one could not have been applied to that of potash, for the ancients do not seem to have distinguished between the two alkalies.

He thinks there is the greater necessity for avoiding any theoretical views in terms because the time is yet far distant for a complete generalisation of chemical facts, and although the antiphlogistic explanation of the phenomena has been uniformly adopted, the motive for employing it has been rather a sense of its beauty and precision than a conviction of its permanency and truth.

> The discovery of the agencies of the gases destroyed the hypothesis of Stahl. The knowledge of the powers and effects of the etherial substances may at a future time possibly act a similar part with regard to the more refined and ingenious hypothesis of Lavoisier; but in the present state of our knowledge, it appears the best approximation that has been made to a perfect logic of chemistry.

Led by analogy, Davy soon convinced himself that the volatile alkali—ammonia—also contained oxygen, and in amount not less than 7 or 8 per cent. It is not necessary to go into detail concerning the experiments on which this erroneous conclusion was founded. Davy was subsequently made aware of his error; but at the time he seemed anxious to overturn—as, indeed, he did in the end, but on other grounds—the Lavoisierian doctrine that oxygen was the principle of acidity, by showing that it was equally the principle of alkalescence.

In concluding his paper, he mentions that he has begun experiments on the alkaline earths.

> From analogy alone it is reasonable to expect that the alkaline earths are compounds of a similar nature to the fixed alkalies, peculiar highly combustible metallic bases united to oxygen. I have tried some experiments upon barytes and strontites, and they go far towards proving that this must be the case.
>
> Barytes and strontites have the strongest relations to the fixed alkalies of any of the earthy bodies; but there is a chain of resemblances through lime, magnesia, glucina, alumina, and silex. And by the agencies of batteries sufficiently strong, and by the application of proper circumstances, there is no small reason to hope that even these refractory bodies will yield their elements to the methods of analysis by electrical attraction and repulsion.

Although certain of the conjectures with which the paper terminates have been proved to be erroneous, others have been shown to be sound. Thus he points out that the metals of the alkalis will undoubtedly prove powerful agents for analysis:

> Having an affinity for oxygen stronger than any other known substances they may possibly supersede the application of electricity to some of the undecompounded bodies.

Such is a brief summary of the contents of one of the most classical papers in the *Philosophical Transactions*. Its publication created an extraordinary sensation, not less profound, and certainly more general from the very nature of the subject, than that which followed his first Bakerian lecture. That potash and soda should contain metals—and such metals!—was undreamt of, and was a shock to the settled convictions of persons who, like the Aberdonian professor, declared that this "and Davy was a very troublesome person in chemistry."

But this "troublesome person" had well nigh ceased from troubling any more. Almost immediately after the delivery of his lecture he collapsed—struck down by an illness which nearly proved fatal, and for weeks his life hung on a thread. He had been in a low feverish condition for some time previously, and a great dread had fallen upon him that he should die before he had completed his discoveries. It was in this condition of body and mind that he applied himself to the task of putting together an account of his results. Four days after this was given to the world he took to his bed, and he remained there for nine weeks. Such a blow following hard on such a triumph, aroused the liveliest sympathy. The doors of the Royal Institution were beset by anxious inquirers. His physicians, Babington, Frank, and Baillie, tended him with the greatest assiduity. Mrs Greenwood, the housekeeper, and his cousin, Edmund Davy, nursed him night and day. So great was the popular feeling that, when he was at the worst, written reports of his condition at various periods of the day had to be posted in the hall. The strength of the feeling may be gleaned, too, from the sentences with which Dr Dibdin began his lecture introductory to the session of 1808:

The Managers of this Institution have requested me to impart to you that intelligence, which no one who is alive to the best feelings of human nature can hear without the mixed emotions of sorrow and delight.

Mr Davy, whose frequent and powerful addresses from this place, supported by his ingenious experiments, have been so long and so well known to you, has for the last five weeks beep struggling between life and death. The effects of these experiments recently made in illustration of his late splendid discovery, added to consequent bodily weakness, brought on a fever so violent as to threaten the extinction of life. Over him it might emphatically be said in the language of our immortal Milton, that

'… Death his dart
Shook, but delayed to strike.'

If it had pleased Providence to deprive the world of all further benefit from his original talents and intense application there has certainly been sufficient already effected by him to entitle him to be classed among the brightest scientific luminaries of his country.

After having given an outline of Davy's investigations "at the particular request of the Managers," Dr Dibdin proceeds

These may justly be placed amongst the most brilliant and valuable discoveries which have ever been made in chemistry, for a great chasm in the chemical system has been filled up; a blaze of light has been diffused over that part which before was utterly dark; and new views have been opened, so numerous and interesting, that the more any man who is versed in chemistry reflects on them, the more he finds to admire and to heighten his expectation of future important results.

Mr Davy's name, in consequence of these discoveries, will be always recorded in the annals of science amongst those of the most illustrious philosophers of his time. His country with reason will be proud of him, and it is no small honour to the Royal Institution that these great discoveries have been made within its walls; in that laboratory, and by those instruments, which from the zeal of promoting useful knowledge have, with so much propriety, been placed at the disposal and for the use of its in most excellent professor of chemistry.

Dr Dibdin then informs his auditors that Davy's illness, severe as it had been, was now beginning to abate, and that it may be reasonably hoped that the period of convalescence was not very remote.

His bodily weakness, however, continued for some time, and it was not until the middle of March that he was able to resume his duties as lecturer. His mind, as his notebooks show, much more quickly recovered its wonted vigour. Perhaps it was in that condition of melancholy and debility produced by sickness, which he regarded as favourable to intellectual exertion, when, as he says, "the mind necessarily becomes contemplative when the body is no longer active, and the empire of sensation yields to that of imagination," that he finished the poem beginning

Lo! o'er the earth the kindling spirits pour
    The flames of life that bounteous Nature gives;
The limpid dew becomes the rosy flower,
    The insensate dust awakes, and moves, and lives.

It is too long to give here, but of all his poetical effusions it is perhaps the best, as it certainly is the most highly-polished.

One proof of what Davy was to the Royal Institution is seen in the position to which it was reduced in consequence of his protracted illness. In the early part of the previous December the Managers made the following announcement:

> Mr Davy, having been confined to his bed this last fortnight by a severe illness, the Managers are under the painful necessity of giving notice that the lectures will not commence until the first week of January next.

By the interruption of the lectures the income of the Institution was greatly diminished; it fell from £4,141 in the preceding year to £1,560. This was the low-water mark of its financial state. How acute was the condition may be seen from the report of the Visitors in 1808.

Davy, although better, was still in bed, confined there by the want of a sofa in his room. This was not provided by the Managers until January 25th, when, as the minutes tell us, they furnished him with one at a cost of three guineas. One would have thought he might have had Albemarle Street blocked with sofas if some of those lady-friends who sent him sonnets, and intrigued for his company at their salons, had only known of his condition.

The laboratory journals show that on April 19th he was able to resume his experiments, and that he proceeded to attack the composition of muriatic [hydro-chloric] acid. The note runs, "Indications of the decomposition of muriatic acid. To use every effort to ensure accuracy in the results." He seems to have decomposed muriatic acid gas by means of charcoal terminals, and also to have acted on a mixture of dry calcium chloride and mercury.

On June 30th he contributed a paper to the Royal Society on "Electro-Chemical Researches on the Decomposition of the Earths; with Observations on the Metals obtained from the alkaline Earths, and on the Amalgam procured from Ammonia." That the earths would turn out to be related to the metal was surmised by Becher and Stahl. Boyle considered it possible that metals might be produced from them, and Neumann described unsuccessful experiments to obtain a metal from quicklime. Bergman imagined that baryta was a metallic calx, and Baron that alumina contained a metal. The supposition that the calces were all compounds of metals was, of course, a part of the antiphlogistic doctrine; but Lavoisier never hazarded any conjecture as to the nature of potash and soda. It went almost without saying therefore that when Davy had demonstrated the real character of the fixed alkalis, the alkaline earths would be found to have an analogous constitution.

The attempts made by Davy to decompose the alkaline earths by methods similar to those adopted in the case of potash or soda were not very successful, and it was only when he had received intimation from Berzelius that they might be procured in the form of amalgams by operating in contact with mercury that he obtained any decisive results. In no case, however, was he able to prepare a pure metal, and his description of the physical properties of the substances he actually procured is exceedingly meagre. He seems to have been satisfied for the moment in demonstrating that

> The evidence for the composition of the alkaline earths is of the same kind as that for the composition of the common metallic oxides; and the principles of their decomposition are precisely similar, the inflammable matters in all cases separating at the negative surface in the voltaic circuit, and the oxygen at the positive surface.
>
> These new substances will demand names; and on the same principles as I have named the bases of the fixed alkalies, potassium and sodium, I shall venture to denominate the metals from the alkaline earths barium, strontium, calcium and magnium; the last of these words is undoubtedly objectionable but magnesium has been already applied to metallic manganese by[Bergman] and would consequently have been an equivocal term.

However, as he states in his "Elements of Chemical Philosophy," "the candid criticisms of some philosophical friends" induced him to subsequently change the name to magnesium.

He next made "Inquiries Relative to the Decomposition of Alumine, Silex, Zircone, and Glucine," but although he made a large number of trials, the results were equivocal.

> "Had I been so fortunate," he says, "as to have obtained more certain evidences on this subject, and to have procured the metallic substances I was in search of, I should have proposed for them the names of silicinm, alumium, zirconium, and glucium."

One of the most interesting sections of the paper relates to the production of a so-called amalgam from ammonia, first obtained by Berzelius and Pontin. This curious substance has been the subject of much investigation, and little doubt is now entertained that it is merely a mercurial froth, as first stated by Daniell—that is, mercury distended by ammonia and hydrogen gases. Davy, however, saw in it the proof of the presence of oxygen in ammonia, and of the existence of what he called "the compound basis" of ammonia. He says:

> The more the properties of the amalgam obtained from ammonia are considered the more extraordinary do they appear. Mercury by combination with about 1/1200 part of its weight of new matter is rendered a solid, yet has its specific gravity diminished from 13 to 3, and it retains all its metallic characters; its colour, lustre, opacity, and conducting powers remaining unimpaired. It is scarcely possible to conceive that a substance which forms with mercury so perfect an amalgam, should not be metallic in its own nature; and on this idea to assist the discussion concerning it, it may be conveniently termed ammonium.

Davy's term "ammonium" is still retained in chemical nomenclature, but there is at present no evidence for the independent existence of such an entity; the so-called ammonium amalgam is certainly no proof.

On December 15th, 1808, he delivered his third Bakerian lecture. It was entitled "An Account of some new analytical Researches on the Nature of certain Bodies, particularly the Alkalies, Phosphorus, Sulphur, Carbonaceous Matter, and the Acids hitherto undecompounded, with some general Observations on Chemical Theory." Although this is one of the longest and most laboured of Davy's papers, it is, perhaps, one of the least satisfactory. It is a record of many experiments with few definite results. Few as these were, they yet paved the way for consequences of the greatest importance. Gay Lussac and Thenard, on the publication of Davy's second Bakerian lecture, succeeded in devising a method by which larger quantities of potassium might be obtained than by the electrolytic process. It consisted in passing molten potash over heated metallic iron and condensing the volatilised potassium in naphtha. On heating potassium in ammonia, they found that hydrogen was obtained together with potash, whence they concluded that potassium was a *hydruret of potash*. This experiment was repeated by Davy; he observed the formation of a substance since known as *potassamide*, and completely disproved the conjecture of the French chemists. His experiments on sulphur, phosphorus, and the various forms of carbon were, however, wholly fallacious, and his conclusions as to the non-elementary nature of these substances were erroneous, and were subsequently corrected by him. His work on the decomposition of boracic acid is, however, accurate, and he has every right to be considered as an independent discoverer, with Thenard, of the element subsequently called by him *boron*. At first Davy was inclined "to consider the boracic basis as metallic in its nature," and to propose for it the name of *boracium.* His experiments with "fluoric acid" were vitiated by the circumstance that he worked with a mixture of hydrofluoric acid and silicon fluoride. Unwittingly he obtained small quantities of silicon, although he failed to recognise the individuality of this substance. Nor were the experiments with muriatic acid more decisive. Incidentally he obtained the two chlorides of phosphorus, but for a time their true nature escaped him, although he gives a fairly accurate description of their main properties.

The paper, although containing an account of much experimental work, was evidently put together in haste; it would have been better for his reputation had he delayed its publication. He seems to have been conscious of its imperfections, and to have sought to strengthen his conclusions by new experiments which he gives in an appendix. These, so far from substantiating his views, increased his doubts, and it is remarkable how he misinterpreted the phenomena he observed. Thus in one series of experiments he obtained considerable quantities of the "alcohol of sulphur of Lampadius," and attempted to ascertain its nature, but his preconceptions as to the non-elementary nature of carbon and sulphur prevented him from recognising that it is a sulphide of carbon.

One explanation of this untoward haste is to be found in the position in which Davy was placed. He simply *hungered* for scientific fame, and his appetite grew by what it fed on. There was at the time the most intense spirit of rivalry between the English and French chemists—it was a phase of the national feeling which actuated the two peoples—and, in spite of his phrases, Davy keenly felt what he considered an intrusion into his own field of work. His illness had thrown him back, and the French chemists had stolen a march on him in the meantime. Moreover, he had Berzelius on his flank. All these circumstances, whilst they impelled him to activity, were unfavourable in a man of Davy's temperament to the incubatory period, "the wambling in the wame" process, which is often needed before the true aspect and meaning of things are perceived; and there is no doubt that the fear of being anticipated urged him to the expression of hypotheses and surmises which at a later and calmer period he regretted and renounced.

But such was his position in England at this period, that a Bakerian lecture seemed to be expected from him at each succeeding session of the Royal Society as a matter of course, and he was always ready to respond to the expectation, even if he did not invariably satisfy it.

On November 16th, 1809, he read his fourth Bakerian lecture. It was "On some new Electrochemnical Researches on various Objects particularly the metallic Bodies, from the Alkalies and Earths, and on some Combinations of Hydrogene." He begins by again drawing attention to the various surmises which had been made respecting the true nature of potassium and sodium. Although these substances had been isolated, and in the hands of chemists for upwards of two years, their properties were so extraordinary when compared with those of the metals in general, that many philosophers hesitated to consider them as true metals. Gay Lussac and Thenard, as already mentioned, regarded them as compounds of potash or soda with hydrogen; Curaudau as combinations of carbon or carbon and hydrogen with the alkalis; whilst an ingenious inquirer in this country communicated to Nicholson's Journal his belief that they were really composed of oxygen and hydrogen! Davy, in the light of the fuller knowledge he obtained from Gay Lussac and Thenard's paper in the *Mem. d'Arcueil*—a copy of which he owed to Berthollet—had no difficulty in again proving "that by the operation of potassium upon ammonia, it is not a *metallic* body that is decompounded, but the volatile alkali, and that the hydrogen produced does not arise from the potassium, as is asserted by the French chemists, but from the *ammonia*."

M. Curaudau's hypothesis is shown to be based upon the accidental association of naphtha with the metals he employed. In repeating some experiments of Ritter's, designed to show that potassium contained hydrogen, Davy was led to the discovery of *telluretted hydrogen*, the properties of which he describes in some detail. Tellurium at that time was regarded as a metal, but Davy points out its strong analogies to sulphur, with which element, indeed, it is now classed. Incidentally he throws light upon the nature of the intolerably fetid product known as "the fuming liquor of Cadet," obtained by distilling acetate of potash with arsenious oxide. On account of its extreme inflammability, it was thought by Davy that this liquid might possibly be a pyrophorus or volatile alloy of potassium and arsenic.

> From a repetition of the process I find that though potash is decompounded in this operation yet that the volatile substance is not an alloy of potassium but contains charcoal and arsenic probably with hydrogen. The gases not absorbable by water given off in this operation are peculiar. Their smell is intensely fetid. They are inflammable, and seem to contain charcoal, arsenic and hydrogen: whether they are mixtures of various gases, or a single compound, I am not at present able to decide.

So far as it goes, this description of the nature of the substance is correct; it was Bunsen, in 1837, who first demonstrated the real character of "the fuming liquor of Cadet."

The paper is noteworthy for the clear distinction which is drawn for the first time between potash hydrate (potassium hydroxide of modern nomenclature) and potassium oxide, the product formed by heating the metal in ordinary oxygen.

There is much in the rest of the paper that is ingenious and suggestive, and not a few isolated facts that seem to have been lost sight of; or rediscovered by subsequent observers, such, for example, as the action of potassium upon metallic iron—an action which has vitiated the attempts to determine the vapour density of that metal in iron vessels. It is curious to note with what persistency Davy clings to the belief that nitrogen will turn out to be a compound substance, and with what pertinacity he importunes it to give up its components. At times he thinks he is on the verge of proof. "I hope on Thursday," he wrote to his friend Children, "to show you nitrogen as a complete wreck, torn to pieces in different ways." But still nitrogen, with that passive immutability which is characteristic of it, in spite of every form of torture, remained whole and indissoluble. On this point he wrote in the Laboratory Journal under date February 15th: "Were a description, indeed, to be given of all the experiments I have made, of all the difficulties I have encountered, of the doubts that have occurred, and the hypotheses formed—." But the sentence was not finished. The attack was renewed and continued throughout the whole of the spring and summer, until, fairly baffled, Davy confessed himself beaten, and turned his attention to other matters. The condition of his laboratory at this time may be gleaned from the following note in the Journal:

Objects much wanted in the laboratory of the Royal Institution: cleanliness, neatness and regularity.

The laboratory must be cleaned every morning when operations are going on before ten o'clock.

It is the business of W. Payne to do this, and it is the duty of Mr E. Davy to see that it is done and to take care of and keep in order the apparatus.

There must be in the laboratory pen, ink, paper, and wafers, and these must not be kept in the slovenly manner in which they are usually kept. I am now writing with a pen and ink such as was never used in any other place.

Then follows a list of articles wanting, "including most of the common metallic and saline solutions."

The laboratory is constantly in a state of dirt and confusion.

There must be a roller with a coarse towel for washing the hands and a basin of water and soap, and every week at least a whole morning must be devoted to the inspection and ordering of the voltaic battery.

It would be interesting to know the comments of the persons named in this note as to the cause of the dirt and confusion which reigned in the laboratory. Davy was perfectly reckless with apparatus; with him to think was to act, and he frequently had half a dozen experiments going on simultaneously, upon disconnected parts of the same inquiry. Anyone who has had the opportunity of seeing his laboratory notes, or of glancing over the rough drafts of his memoirs, which have been preserved by the pious care of Faraday, will appreciate the significance of the remarks upon his writing materials. His usual method of erasure was by dipping his finger in the ink-pot; and, if we may be pardoned the use of the colloquialism, he was simply "Death on pens!"

1.   It is frequently stated that Davy was enabled to isolate the metals of the alkalis because of the *large* and powerful voltaic battery which he had at his disposal in the Royal Institution. This is not correct. The battery he employed was of very moderate dimensions, and not by any means extraordinary in power. It was the success he thus achieved that caused the large battery, which is probably referred to, to be constructed, by special subscription, in 1809.

VII

CHLORINE

THE RIVALRY BETWEEN THE FRENCH and English chemists continued, but. it took a new departure. Gay Lussac and Thenard had stolen a march on Davy by their discovery of a chemical method of making the metals of the alkalis, whereby they were able to use these metals as chemical reagents to greater advantage; but the tables were quickly turned. On July 12th, 1810, Davy read to the Royal Society his memorable paper "On the oxymuriatic Acid, its Nature and Combinations; and on the Elements of the muriatic Acid: with some Experiments On Sulphur and Phosphorus, made in the Laboratory of the Royal Institution." This paper, in which he first demonstrates the nature of chlorine, is very short—only some twenty-six quarto pages—but it is unquestionably one of the most brilliant, as it is one of the most forcible of his productions.

Davy is here seen at his best. He is bold and yet wary, and as dexterous as trenchant; so confident is he in the strength of his position that he casts aside every argument that might tell in his favour, unless it is based on the most unimpeachable evidence. It is difficult to know what to admire most—the clearness of perception, the precision of the statement, the strictness of the logic, the aptness of the illustration, or the argumentative skill with which the whole is marshalled and presented. As a piece of induction, the memoir is a model of its kind, and as an exercise in "the scientific use of the imagination" it has few equals. Most scientific papers will stand a considerable amount of winnowing, and there is no assay-master more scrupulously strict than Time. "The more a science advances, the more it becomes concentrated in little books," says Leibnitz; but the most fastidious of critics might read and re-read this work without wishing to omit or amend a sentence.

Every chemical student today is told that the elementary nature of chlorine was first *demonstrated* by Davy, and if the student is informed what Davy meant by the term "element," the statement is not incorrect. What, however, Davy actually did was to demonstrate that the substance called oxymuriatic acid contained no oxygen; that it was a peculiar substance which "has not as yet been decompounded," and therefore is "elementary as far as our knowledge extends." The very character of the name which he suggested indicates this cautious and philosophical view. In making the suggestion, he says:

> To call a body which is not known to contain oxygen and which cannot contain muriatic acid, oxymuriatic acid, is contrary to the principles of that nomenclature in which it is adopted; and an alteration of it seems necessary to assist the progress of discussion, and to diffuse just ideas on the subject. If the great discoverer of this substance [Schleele, who first observed it in 1774] had signified it by any simple name, it would have been proper to have recurred to it; but, *dephlogisticated marine acid* is a term which can hardly be adopted in the present advanced era of the science.
>
> After consulting some of the most eminent chemical philosophers in this country, it has been judged most proper to suggest a name founded upon one of its obvious and characteristic properties—its colour, and to call it *chlorine*, or *chloric* gas.[1]
>
> Should it hereafter be discovered to be compound, and even to contain oxygen, this name can imply no error, and cannot necessarily require a change.

As the actual facts and arguments on which Davy based his views are seldom set forth in textbooks, or presented to the student by teachers, it may be desirable to give a detailed account of his famous memoir. He begins by saying:

> The illustrious discoverer of the oxymuriatic acid considered it as muriatic acid freed from hydrogen; and the common muriatic acid as a compound of hydrogen and oxymuriatic acid; and on this theory he denominated oxymuriatic acid dephlogisticated muriatic acid.
>
> M. Berthollet, a few years after the discovery of Scheele, made a number of important and curious experiments on this body; from which he concluded that it was composed of muriatic acid and oxygen; and this idea for nearly twenty years has been almost universally adopted.

Having thus accurately stated the position, he proceeds to attack it. In the first place, he points out that Henry, ten years before, had shown that hydrogen could be produced from muriatic acid gas by the agency of electricity; this hydrogen was assumed by Henry to be due to water contained in the gas. Davy, in his Bakerian lecture of 1808, had shown that muriatic acid gas gave hydrogen when treated with potassium, and he had stated "that muriatic acid can in no instance be procured from oxymuriatic gas, or from dry muriates, unless water or its elements be present."

Gay Lussac and Thenard had concluded "that muriatic acid gas contains about one-quarter of its weight of water; and that oxymuriatic acid is not *decomposable* by any substances but hydrogen, or such as can form triple combinations with it."

He then points out, what he had already stated in a former paper, that charcoal freed from hydrogen and moisture by intense ignition *in vacuo* may be heated to whiteness by the voltaic battery in oxymuriatic or muriatic acid gases without affecting any change in them.

It now occurred to him that if the liquor of Libavius (stannic chloride) is a combination of muriatic acid and oxide of tin, as then surmised, oxide of tin ought to be separated from it by means of ammonia. On admitting ammonia gas to the tin chloride over mercury, the substances combined with great heat, a white solid was obtained; "some of it was heated to ascertain if it contained oxide of tin, but the whole volatilised, producing dense pungent fumes." The experiment was repeated with every care, but no oxide of tin could be obtained.

He was next led to study the behaviour of ammonia with the substances he had formerly obtained, by the action of oxymuriatic gas on phosphorus (see p. 105). One of these is solid, and is now known as phosphorus pentachloride; the other is liquid, and is termed phosphorus trichloride.

"The first," he says, "on the generally received theory of the nature of oxymuriatic acid, must be considered as a compound of muriatic acid and phosphoric acid. It occurred to me that if the acids of phosphorus really existed in these combinations, it would not be difficult to obtain them, and thus to gain proof of the existence of oxygen in oxymuriatic acid."

He therefore brought ammonia gas into contact with the solid compound of oxymuriatic acid and phosphorus. Much heat was produced, and a white opaque powder was formed.

Supposing that this substance was composed of the dry muriate and phosphate of ammonia; as muriate of ammonia is very volatile, and as ammonia is driven off from phosphoric acid, by a heat below redness I conceived that by igniting the product obtained I should procure phosphoric acid … but found to my great surprise that it was not at all volatile nor decomposable at this degree of heat, and that it gave off no gaseous matter. The circumstance that a substance composed principally of oxymuriatic acid and ammonia should resist decomposition or change at so high a temperature induced me to pay particular attention to the properties of this new body.

What he actually obtained was mainly a mixture of the so-called *phospham* and *chlorophosphamide*, remarkably stable substances, the characteristic properties of which he describes with accuracy. He then examined the action of ammonia gas on sulphur chloride, "the sulphuretted muriatic liquor of Dr Thomson," but as the compounds formed

did not present the same uniform and interesting properties as that from the phosphoric sublimate, I did not examine them minutely: I contented myself by ascertaining that no substance known to contain oxygen could be procured from oxymuriatic acid in this mode of operation.

He then shows that ammonia and oxymuriatic acid, in condensing to sal ammoniac with liberation of nitrogen, contrary to the general belief, form no water. According to Cruickshank, who appears to have been the first to make the observation, "hydrogenous gas" required rather more than its own volume of oxygenated muriatic acid to saturate it when a mixture of the two was exploded by means of the electric spark, "the products being water and muriatic acid." Gay Lussac and Thenard had stated that no water was thus formed.

"I have attempted," says Davy, "to make the experiment still more refined by drying the oxymuriatic acid and the hydrogen by introducing them into vessels containing muriate of lime [calcium chloride] and by suffering them to combine at common temperatures; but I have never been able to avoid a slight condensation; though in proportion as the gases were free from oxygen or water, this condensation diminished.[2]

"MM. Gay Lussac and Thenard have proved by a copious collection of instances, that in the usual cases where oxygen is procured from oxymuriatic acid, water is always present, and muriatic acid gas is for now as it is shewn that oxymuriatic acid gas is converted into muriatic acid gas by combining with hydrogen, it is scarcely possible to avoid the conclusion, that the oxygen is derived from the decomposition of the water, and consequently that the idea of the existence of water in muriatic acid gas, is hypothetical, depending upon an assumption which has not yet been proved—the existence of oxygen in oxymuriatic acid gas.

"MM. Gay Lussac and Thenard indeed have stated an experiment, which they consider as proving that muriatic acid gas contains one quarter of its weight of combined water. They passed this gas over litharge, and obtained so much water; but it is obvious, that in this case, they formed the same compound as that produced by the action of oxymuriatic acid on lead; and in this process the muriatic acid must lose its hydrogen and the lead its oxygen; which of course would form water; these able chemists, indeed, from the conclusion of their memoir, seem aware, that such an explanation may be given, for they say, that the oxymuriatic acid *may* be considered as a simple body."

He then repeats the experiments which first led him to suspect the existence of combined water in muriatic acid.

When mercury is made to act upon 1 volume of muriatic acid gas, by voltaic electricity, all the acid disappears, calomel is formed, and about ·5 of hydrogen evolved.

The same result is obtained by the use of potassium.

And in some experiments made very carefully by my brother, Mr John Davy, on the decomposition of muriatic acid gas, by heated tin and zinc, hydrogen, equal to about half its volume, was disengaged, and metallic muriates, the same as those produced by the combustion of tin and zinc in oxymuriatic gas, resulted.

It is evident from this series of observations, that Scheele's view (though obscured by terms derived from a vague and unfounded general theory) of the nature of the oxymuriatic and muriatic acids, may be considered as an expression of facts; whilst the view adopted by the French school of chemistry, and which, till it is minutely examined, appears so beautiful and satisfactory rests in the Present state of our knowledge upon hypothetical grounds.

He then proceeds to explain the action of water upon the chlorides of tin, and phosphorus; and shows that it is by the decomposition of the water that the hydrogen is furnished to the oxymuriatic acid, and the oxygen to the tin and phosphorus.

The vivid combustion of bodies in oxymuriatic acid gas, at first view, appears a reason why oxygen should be admitted in it; but heat and light are merely results of the intense agency of combination. Sulphur and metals, alkaline earths and acids become ignited during their mutual agency; and such an effect might be expected in an operation so rapid as that of oxymuriatic acid upon metals and inflammable bodies.

That the quantity of hydrogen evolved during the decomposition of muriatic acid gas by metals, is the same that would be produced during the decomposition of water by the same bodies, appears, at first view, an evidence in favour of the existence of water in muriatic acid gas; but as there is only one known combination of hydrogen with oxymuriatic acid, one quantity must always be separated. Hydrogen is disengaged from its oxymuriatic combination by a metal, in the same manner as one metal is disengaged by another from similar combinations.

He once more shows that by the strongest analytical power he can command oxymuriatic acid fails to yield any substance differing from itself:

I have caused strong explosions from an electrical jar, to pass through oxymuriatic gas, by means of points of platina, for several hours in succession; but it seemed not to undergo the slightest change.

Such, then, are the reasons which induced Davy to consider that oxymuriatic acid contains no oxygen; that it had hitherto been "undecompounded," and that, therefore, by the strict logic of chemistry, it was to be regarded as an elementary body. Had his paper concluded at this point, his position would have been unassailable, even in the light of nearly ninety years of subsequent work. But he could not stop here. Berthollet, the author of the prevailing theory, had discovered a salt then known as *hyper-oxymuriate* of potash, presumably capable of furnishing an acid termed by Chenevix *hyper-oxygenised muriatic acid*. This salt is now termed potassium chlorate, after the acid which Davy subsequently succeeded in isolating, and which, when the chlorine theory was generally accepted, was called chloric acid by Gay Lussac. The existence of the hyper-oxymuriate of potash was for a time a stumbling-block, and Davy sought to explain it on the assumption that it was nothing more than a triple compound of oxymuriatic acid, potassium, and oxygen.

We have no right to assume the existence of any peculiar acid in it, or of a considerable portion of combined water; and it is perhaps more conformable to the analogy of chemistry to suppose the large quantity of oxygen combined with the potassium, which we know has an intense affinity for oxygen, and which from some experiments, I am inclined to believe, is capable of combining directly with more oxygen than exists in potash, than with the oxymuriatic acid which, as far as is known, has no affinity for that substance.

It is perfectly true, as Davy surmised, that potassium can combine with more oxygen than is contained in potash, but it is no less true, as he himself proved by his discovery of the so-called *euchlorine*, that chlorine can combine with oxygen. Although he made several attempts to isolate Mr Chenevix's hyper-oxygenised muriatic acid, he was not successful at the time, and was evidently disposed to doubt its separate existence.

The remaining portion of the paper, although of interest as exemplifying Davy's power of dealing with the broad issues which his views raise, need not detain us now. He seizes the opportunity, however, to correct his statements with regard to the presumed compound nature of sulphur and phosphorus, and gives details of observations, some of which, as in other of his papers, have been "discovered" by subsequent observers. Thus he states:

> I have never been able to burn sulphur in oxygen without forming sulphuric acid in small quantities; but in several experiments I have obtained from 92 to 98 parts of sulphurous acid from 100 of oxygen in volume; from which I am inclined to believe that sulphurous acid consists of sulphur dissolved in an equal volume of oxygen.

It was hardly to be expected that views so entirely opposed to the convictions of chemists at the time should pass unchallenged. Berzelius, the countryman of Scheele, warmly defended the doctrine of the French School, and yet another Scotch professor sought to show that Davy was still "very troublesome." The controversy, in which Davy himself took little part, occasioned considerable stir at the period, and was even of interest outside philosophical circles. The discussion was not without its uses, inasmuch as it led to fresh discoveries. The noise of it all, however, is now forgotten. Berzelius eventually enjoined his cook to speak no longer of oxymuriatic acid "Thou must call it *chlorine*, Anna; that is better." Dr Murray, with the pertinacity of his race, still clung to the old doctrine, and defended it with no little dialectical subtlety, but he alone was faithful among the faithless. It is true there has been an occasional flutter in the dovecots since these times, and the faith of chemists in the validity of Davy's teaching has been once or twice assailed, but as yet it has survived all assaults.

The Royal Institution possesses a book which no lover of science can regard with other than reverential interest. It is a small, well-bound quarto of some 386 manuscript pages of notes taken by Michael Faraday, when a bookbinder's apprentice, of the last of Davy's lectures at the Institution. A Mr Dance—his name deserves to be held in remembrance—had given the youth a ticket for the lectures, and Faraday, perched in the gallery over the clock, had zealously followed the expositions of the brilliant lecturer, and had subsequently, when asking for an engagement at the Institution, sent in these notes, neatly written out and embellished with drawings of the apparatus, to the Professor as evidence of the applicant's "knowledge, diligence and order." Among the lectures is one on chlorine, given on March 14th, 1812, the notes of which are as characteristic of the auditor as of the lecturer. We read:

> Accustomed for years to consider the chemical principles of the French School of Physical Sciences as correct, I had adopted them and put faith in them until they became prejudices, and I even felt unwilling to give them up when my judgment was fully convinced by experiment that they were erroneous. I know that this is the case in some degree with almost every person; he is unwilling to believe that he is wrong and therefore feels averse to adopt what is right when it opposes his principles.

Then follows an account of various experiments showing the properties of chlorine, and the proofs that it contains no oxygen:

> Oxygen does combine with chlorine. I have ventured to name the compound *euchlorine;* it is of a very bright yellow-green colour. Names should represent things not opinions for in the last case they often tend to misrepresent and mislead.
>
> Had Mr Berthollet obtained oxygen from chlorine there would have been no error in his theory, but by not attending to the minute circumstances of his experiment, by not ascertaining that the water present acted no part and was not decomposed he fell into an error, and of course all the conclusions he drew were false and erroneous. Nothing should be allowed but what can be proved by experiment, and nothing should be taken for granted upon analogy or supposition.

Faraday concludes as follows:

> Mr Davy now proceeded to comment and make observations on the former theory of chlorine gas. Here I was unable to follow him. The plan which I

pursue in taking of notes is convenient and self-sufficient with respect to the theoretical and also the practical part of the lecture, but for the embellishments and ornaments of it it will not answer. Mr Davy's language at those times is so superior (and indeed throughout the whole course of the lecture) that then I am infinitely below him, and am incapable of following him even in an humble style. Therefore I shall not attempt it; it will be sufficient to give a kind of contents of it. He said that hypotheses should not be considered as facts and built upon accordingly. Nevertheless, if cautiously pursued, they might lead to mature fruit. That nothing should be taken for granted unless proved. By considering oxygen as contained in chlorine the whole chemical world had been wrapped in error respecting that body for more than one-third of a century.

He noticed that all the truly great scientific men were possessed of great humility and diffidence of their own opinions and powers. He spoke of Scheele, the discoverer of chlorine; observed that he possessed a truly philosophical spirit, gave up his opinions when he supposed them to be erroneous, and without hesitation or reluctance adopted those of others which he considered more correct; admired his spirit and recommended it to all philosophers; compared it to corn, which looked but simple and insignificant in blossom, and asked for little praise, yet was the support of man.

In his fifth Bakerian lecture, "On some of the Combinations of Oxymuriatic Gas and Oxygene, and on the chemical Relations of these Principles to inflammable Bodies," read before the Royal Society on November 15th, 1810, he still further developed his ideas respecting the nature of chlorine. Gay Lussac and Thenard, who had convinced themselves that potassium and sodium are not hydrates of potash and soda, had made known the fact that potassium can combine with oxygen in more than one proportion; and Davy had confirmed their conclusion, seeing in it a further proof of his views concerning the constitution of the hyper-oxymuriate of potash. He then studied the behaviour of a large number of the metals and their oxides with chlorine, making in many cases quantitative determinations, from which very fair approximations to the combining proportions or atomic weights of the substances may be deduced. Thus, he says "the number representing the proportion in which mercury combines must be about 200," and that "the quantity of chlorine in corrosive sublimate is exactly double that in calomel, and that the orange oxide contains twice as much oxygen as the black, the mercury being considered the same in all." The atomic weight of silver deducible from the amount of chlorine taken up by that metal during its conversion into horn silver is almost exactly the value obtained by the most rigorous analyses of modern times. It is, however, noteworthy that in this paper Davy is brought into sharp conflict with Dalton, and there is a characteristic exhibition of temper in the way in which he protests against the manner in which Dalton had sought to use certain of his numerical estimations in deducing the weights of atoms. The comparative merits of Mr Higgins and John Dalton as the real authors of the explanation of the laws of chemical combination have now been fully and finally assessed, but it was wholly unnecessary for the purpose of Davy's contention to underrate the originality of the Manchester chemist. Dalton was no doubt wrong in the assumption that 47 represented the weight of the atom of nitrogen, and Davy was right in pointing out the invalidity of the basis on which this assumption rested, and in his statement that 13·4 more nearly represented the smallest proportion in which nitrogen is known to combine. Davy says:

> I shall enter no further at present into an examination of the opinions, results, and conclusions of my learned friend; I am however obliged to dissent from most of them, and to protest against the interpretations that he has been pleased to make of my experiments; and I trust to his judgment and candour for a correction of his views.
>
> It is impossible not to admire the ingenuity and talent with which Mr Dalton has arranged, combined, weighed, measured, and figured his atoms; but it is not, I conceive, on any speculations upon the ultimate particles of matter, that the true theory of definite proportions must ultimately rest. It has a surer basis in the mutual decomposition of the neutral salts, observed by Richter and Guyton de Morveau, in the mutual decompositions of the compounds of hydrogen and nitrogen, of nitrogen and oxygen, of water and the oxymuriatic compounds; in the multiples of oxygen in the nitrous compounds ; and those of acids in salts, observed by Drs. Wollaston and Thomson; and above all, in the decompositions by the Voltaic apparatus, where oxygen and hydrogen, oxygen and inflammable bodies, acids and alkalies, &c., must separate in uniform ratios.

It has been alleged that Davy in thus expressing himself offered a kind of factious opposition to the views of Dalton. In so far as they were *atomic*, this is possibly true, for Davy never brought himself to regard the fact of chemical combination occurring in definite proportions as admitting of the simple mechanical explanation of Dalton, which he considered too speculative. That, however, he did ample justice to Dalton's merits ultimately will be seen from the terms in which he speaks of them on the occasion of the award to Dalton in 1826 of the first of the Royal medals. In one of his unfinished Dialogues, written shortly before his death, "On the Powers which act upon Matter and produce Chemical Changes," he thus expresses himself:

> The atomic doctrine, or theory, has been embraced by several modern chemists; but the development of it is owing to Mr Dalton who seems to have been the first person to generalize the facts, of chemistry relating to definite proportions … Mr W. Higgins appears to have had only some loose idea of particles combining with particles, without any profound views of the quantity being unalterable; and there is good reason for thinking that these ideas, as he expresses them, were gained from another source, Dr Bryan Higgins, who many years before supported the notion, that chemical substances were formed of molecules, either simple or compound, surrounded by an atmosphere of heat; and his views, though not developed with precision, approached nearer to those of Mr Dalton, than those of his cousin. But neither of these gentlemen attempted any statical expressions; and to Richter and Dalton belongs the exclusive merit of having made the doctrine practicable. As a theoretical view, other authors have a claim to it, and the early followers of Newton, such as Kiel, Hartley, and Marzucchi, all attempted a corpuscular chemistry, founded upon figure, weight, and attractive power of the ultimate particles of matter; but this chemistry was of no real use, and had no other foundation than in the imagination. Indeed, in my opinion, Mr Dalton is too much of an *Atomic Philosopher*; and in making atoms arrange themselves according to his own hypothesis, he has often indulged in vain speculation; and the essential and truly useful part of his doctrine, the expression of the quantities in which bodies combine, is perfectly independent of any views respecting the ultimate nature either of matter or its elements.

He concludes the paper in which he so minutely studied the action of chlorine upon oxides by asking, if it be said that the oxygen arises from the decomposition of the oxymuriatic gas and not from the oxides, why is it always the quantity contained in the oxide that is evolved? And why in some cases, as those of the peroxides of potassium and sodium, it bears no relation to the quantity of oxymuriatic gas?

> When potassium is burnt in oxymuriatic gas, a dry compound is obtained. If potassium combined with oxygen is employed, the whole of the oxygen is expelled, and the same compound formed. It is contrary to sound logic to say, that this exact quantity of oxygen is given off from a body not known to be compound, when we are certain of its existence in another; and all the cases are parallel.

An argument in favour of the existence of oxygen in chlorine might be derived from the circumstance of the formation of the latter gas by the action of muriatic acid on peroxides. Davy found that, by heating muriatic acid gas in contact with dry peroxide of manganese, water was rapidly formed and oxymuriatic gas produced.

> Now as muriatic acid gas is known to consist of oxymuriatic gas and hydrogen, there is no simple explanation of the result, except by saying that the hydrogen of the muriatic acid combined with oxygen from the peroxide to produce water.

The bleaching power of chlorine had been explained by Scheele on the supposition that it destroyed colours by combining with phlogiston. Berthollet considered it to act by supplying oxygen. Davy then made the well-known experiment proving that the dry gas "is incapable of altering vegetable colours, and that its operation in bleaching depends entirely upon its property of decomposing water and liberating its oxygen." It had been supposed that oxymnuriatic acid gas was capable of being condensed and crystallised at a low temperature. He shows that it was only damp chlorine or its solution in water that yielded any solid product. He exposed the pure gas, dried by muriate of lime, to a temperature of 40° F., without observing any change. It is curious, however, that liquid chlorine had actually been obtained by Northmore five years before by heating the so-called hydrate of chlorine under pressure. The phenomenon was misunderstood, and it was reserved for Faraday, in 1823, to show that the product was actually the liquefied gas.

Davy, who was not always happy in his suggestions as to chemical nomenclature, proposed to denote the compounds of oxymuriatic gas by the names of their bases with the termination *ane*.

> Thus, argentane may signify horn-silver; stannane Libavius's liquor; antimonane, butter of antimony; sulphurane, Dr Thomson's sulphurated liquor, and so on for the rest … In cases when two or more proportions of inflammable matter combine with one of gas; or two or more of gas with one of inflammable matter, it may be convenient to signify the proportions by affixing vowels before the name, when the inflammable matter predominates, and after the name when the gas is in excess; and in the order of the alphabet, *a* signifying two, *e*, three, *i*, four and so on.

Thus he called phosphorus pentachloride *phos phorana*, and the trichloride *phosphorane*, because there was a larger percentage proportion of phosphorus in the latter compound than in the former. That Davy was not unaware of the difficulties and inconveniences of such a system of nomenclature may be inferred from what he says in his "Elements" concerning the names for the two chlorides of mercury, the true composition of which he was the first to discover:

> The names *mercurane* and *mercurana* which may be adopted to signify the relations of their composition, are too similar to each other to be safely used as familiar appellations for the two substances, as corrosive sublimate is a powerful poison, calomel an excellent medicine.

In matters of chemical nomenclature Davy was a great latitudinarian. All that he contended for was that names should be independent of all speculative views, and should rather be derived from some simple and invariable property. It is remarkable, however, that he who invented the happy term "chlorine" should have objected to the word "cyanogen." At the close of the short paper "On the Prussic Basis and Acid," in which he first made known the existence of the cyanides of phosphorus and of iodine, he said:

> I wish M. Gay Lussac could be prevailed upon to give up the inexpressive and difficult names of cyanogen and hydrocyanic acid, and to adopt the simple ones of prussic gas and prussic acid.

By treating the potassium hyper-oxymuriate of Berthollet (potassium chlorate) with hydrochloric acid, a greenish-yellow explosive gas is obtained which Chenevix had referred to as "hyper-oxygenised muriatic acid," and as indicating the existence of a compound of oxymuriatic gas and oxygen in a separate state. Davy, as we have seen, was at first inclined to doubt the existence of this substance, and to consider the gas as simply chlorine. But on comparing it with chlorine prepared in other ways he perceived a difference; its solution in water was of lemon yellow or orange colour; when treated with mercury it becomes of a brilliant yellow green. It is, moreover, highly explosive, especially when heated, even at the warmth of the hand, when it loses its vivid colour, and is resolved into a mixture of oxygen and chlorine. Metals, arsenic, phosphorus, charcoal, nitric oxide, act upon it in a manner different from that of chlorine. Davy makes use of these differences as a proof of the correctness of his views of the nature of chlorine.

> If the power of bodies to burn in oxymuriatic gas depended upon the presence of oxygen, they all ought to burn with much more energy in the new compound; but copper and antimony, and mercury and arsenic and iron and

sulphur have no action upon it, till it is decomposed; and they act then according to their relative attractions on the oxygen, or on the oxymuriatic gas. There is a simple experiment which illustrates this idea. Let a glass vessel containing brass foil be exhausted, and the new gas admitted, no action will take place; throw in a little nitrous gas [nitric oxide], a rapid decomposition occurs, and the metal burns with great brilliancy.

As the new compound in its purest form is possessed of a bright yellow-green colour, it may be expedient to designate it by a name expressive of this circumstance and its relation to oxy-muriatic gas. As I have named that elastic fluid Chlorine; so I venture to propose for this substance the name Euchlorine, or Euchloric gas from ευ and χλωρος. The point of nomenclature I am not inclined to dwell upon. I shall be content to adopt any name that may be considered as most appropriate by the able chemical philosophers attached to this Society" [the Royal Society].

Euchlorine was subsequently discovered by Soubeiran to be a mixture of chlorine and chlorine peroxide, a gas which Davy himself afterwards isolated in a pure state. It is however obvious from the accounts he gives that even in his first paper he must have been experimenting with a fairly pure product, due probably to the circumstance that he had collected the mixed gases over mercury, which retains the greater part of the chlorine. Former experimenters had collected the gas over water, which dissolves the chlorine peroxide more readily than the chlorine. Madame de Staël once observed that an interesting book might be written on the important consequences which have sprung from little differences. It ought to be noted, however, that Davy had himself doubts whether his euchlorine was not a mixture of chlorine and the gas which he subsequently discovered, and to which he says: "I shall not propose to give any name till it is determined whether euchlorine is a mixture or a definite compound."

It has been stated that Davy discovered the two chlorides of phosphorus. In a paper read to the Royal Society on June 18th, 1812, "On some Combinations of Phosphorus and Sulphur and on some other Subjects of Chemical Inquiry," he reverts to these substances, as they "offer decided evidences in favour of an idea that has been for some time prevalent among many enlightened chemists and which I have defended in former papers published in the Philosophical Transactions; namely that bodies unite in definite proportions, and that there is a relation between the quantities in which the same element unites with different elements."

He first makes a determination, singularly accurate for the time of the amount of chlorine contained in the lower chloride, and finds that 13·6 grains on decomposition with water afforded 43 grains of horn-silver; theory requires 42·6 grains. By synthetical experiments he came to the conclusion that the amount of chlorine absorbed by phosphorus to form the higher chloride was exactly double that contained in the lower chloride: he found that 3 grains of phosphorus combined with 20 grains of chlorine: in reality it should require only 17¾ grains.

He shows that by treatment with water the lower chloride yields *phosphorous acid*, the properties and mode of decomposition of which by heat he accurately describes. He further concludes, as the logical consequence of his view of the composition of the two chlorides, and the mode of their decomposition by water, that phosphorous acid contains half the amount of oxygen present in phosphoric acid, the quantity of phosphorus being the same. It is noteworthy that in his argument, as indeed on all subsequent occasions when he speaks of the decomposition of water in definite proportions, he regards water as composed of 2 combining proportions of hydrogen and 1 of oxygen, and the number representing it as 17, oxygen being regarded as 15. Certain of his statements considered in the light of subsequent work are interesting. Thus he says:

> A solid acid volatile at a moderate degree of heat, may be produced by burning phosphorus in very rare air, and this seems to be phosphorous acid free from water; but some phosphoric acid, and some yellow oxide of phosphorus are always formed at the same time.

He also observes that unless the product of the combustion of phosphorus is strongly heated in oxygen it contains phosphorous acid as well as phosphoric acid. He further states that sulphurous acid (sulphur dioxide) consists of equal weights of oxygen and sulphur, which is almost strictly true, and that sulphuretted hydrogen is composed of 1 combining proportion of sulphur and 2 of hydrogen, although his values for the combining proportions of sulphur and oxygen are incorrect. He repeats Dalton's experiment of the formation of "solid sulphuric acid" by the mutual action of sulphur dioxide and nitric oxide, and shows that the substance is only produced in presence of vapour of water; the two substances, he says, then "form a solid crystalline hydrat; which when thrown into water gives off nitrous gas and forms a solution of sulphuric acid." This substance is the so-called "leaden-chamber crystal," or nitro-sulphonic acid, the existence of which was first made known by Scheele.

Davy's conclusions concerning the composition of the oxides and chlorides of phosphorus were subsequently contested by Berzelius and Dulong, who showed that although the amount of chlorine in the lower chloride was identical with that which he had found, the ratio of this amount to that in the higher chloride was as 3 to 5, and not as 1 to 2, and that the same ratio held good as regards the oxygen in phosphorous oxide and phosphoric oxide. Davy, six years afterwards, repeated his experiments, but without discovering the fallacy in his first observations.

The other incidents in Davy's scientific career may be most conveniently dealt with in connection with his personal history.

1. From χλωρος.

2. Theoretically, there should be no contraction. One volume of chlorine combines with one volume of hydrogen to form two volumes of hydrogen chloride [muriatic acid gas]. Dalton's law of gaseous volumes had been established by Gay Lussac before 1810.

VIII

MARRIAGE—KNIGHTHOOD—ELEMENTS OF CHEMICAL PHILOSOPHY— NITROGEN TRICHOLORIDE—FLUORINE

AVY WAS NOW (1810) THIRTY-two years of age, and near the summit of his scientific fame, and perhaps also, says his brother John, who was then in daily association with him, at the height of his happiness.

> He had earned an unsullied and noble reputation; he was loved and admired by friends, who had cheered him on in his career; he had hardly passed the prime of manhood; he was in possession of excellent health; he had open to him almost every source of ordinary recreation and enjoyment; and he had, besides, the unfailing pleasures derived from the active and successful pursuit of science. His letters written at this time, [to his mother and sisters] strongly mark a happy contentment, as well as a very amiable and affectionate state of mind.

His popularity at the Royal Institution was unbounded; indeed, he was the very prop of its existence, and was so recognised. But honourable as his position was, it brought him little more than a competency; and however generously disposed the Managers might have felt towards him, the financial circumstances of the Institution afforded no certainty of a future independence. The Bishop of Durham and Sir Thomas Bernard sought to induce him to enter the Church, in the hope that his talents and eloquence would minister no less to the cause of religion than to his own prospects of preferment. At this period he had serious thoughts of again applying himself to the study of medicine, with a view of practising as a physician, and he actually entered his name at Cambridge and kept some terms there. But whether the unfortunate experience of his colleagues Wollaston and Young deterred him, or whether, as is more probable, Science had too strong a hold upon his affections, it is certain he made no resolute attempt to abandon her.

Money was never an object with Davy, except as the means of procuring him the advantages which the moneyed classes can command; had he cared for it, his talents were a marketable commodity, and would have brought him riches in many ways. The smiling goddess now showed him one way as honourable as it was lucrative and pleasurable. The Dublin Society invited him to lecture to them on the discoveries which had made him famous, with the promise of a more substantial token of their appreciation than the sound of their applause.

The following minutes from the Proceedings of the Society serve to explain this:

> *May* 3, 1810. *Resolved*—That it is the wish of the Society to communicate to the Irish public in the most extended manner (consistent with the engagements of the Society), the knowledge of a science so intimately connected with the improvement of agriculture and the arts, which is their great object to promote; and that, with this view, it appears to them extremely desirable to obtain the fullest communication of the recent discoveries in electro-chemical science which have been made by Mr Davy.
>
> *Resolved*—That application he made to the Royal Society requesting that they be pleased to dispense with the engagements of Mr Davy [as Secretary], so far as to allow the Dublin Society to solicit the favour of his delivering a course of electro-chemical lectures in their new laboratory, as soon as may be convenient after the present course of chemical lectures shall have been completed by their professor, Mr Higgins.
>
> *Resolved*—That the sum of 400 guineas be appropriated out of the funds of the Society, to be presented to Mr Davy, as a remuneration, which they propose him to accept, and as a mark of the importance they attach to the communication they solicit.

We further read: "Mr Davy arrived in Dublin and delivered his course of lectures to a crowded auditory." At the close of his lectures the following resolution was passed:

> *November 29th,* 1810. *Resolved*—That the thanks of the Society be communicated to Mr Professor Davy, for the excellent course of lectures which, at their request, he has delivered in their new laboratory; and to assure him, that the views which led the Society to seek for these communications have been answered even beyond their hopes; that the manner in which he has unfolded his discoveries has not only imparted new and valuable information, but, fur-

ther, appears to have given a direction of the public mind towards chemical and philosophical inquiries, which cannot fail in its consequences to produce the improvement of the sciences, arts, and manufactures in Ireland. That Mr Davy be requested to accept the sum of five hundred guineas from the Society.

From Mr Hare's *Life and Letters of Maria Edgeworth* we gain some further information of the manner in which these lectures were received. In a letter to her cousin, Miss Buxton, Miss Edgeworth writes:

We are to set out for Dublin on the 13th [*November*] to hear Davy's lectures.

Mrs Edgeworth adds:

We spent a few weeks in Dublin. Davy's lectures not only opened a new world of knowledge to ourselves and to our young people, but were especially gratifying to Mr Edgeworth and Maria, confirming, by the eloquence, ingenuity, and philosophy which they displayed, the high idea which they had so early formed of Mr Davy's powers.

Additional evidence of his success is seen in the circumstance that the Society decided to repeat their invitation:

June 13th, 1811. *Resolved*—That a letter be written to Mr Professor Davy requesting him to favour the Dublin Society and the Irish public with a further communication of the recent discoveries in chemical philosophy, and to deliver a course of lectures in their laboratory for that purpose, in the months of November and December next; and requesting that he will also repeat to them, at the same time, the course of lectures in geological science which he has read this year to the Royal Institution; and that he will be so good as to procure for the Society copies of as many of the geological sketches referred to in that course as he may think necessary for the elucidation of the subject; and further requesting him to superintend the construction of a voltaic battery of large plates, for the use of the Society, to be transmitted to them in time for these lectures.

We next read:

December 5th, 1811. *Resolved unanimously*—That the thanks of the Society be communicated to Mr Davy, for the two excellent courses of lectures in chemical and geological science which, at their request, he has delivered in their laboratory, full of valuable information; and which have not merely continued, but materially increased, the spirit of philosophical research in Ireland.

Resolved unanimously—That Mr Davy be requested to accept the sum of £750 as a remuneration on the part of the Society.

On the occasion of his second visit Trinity College, Dublin, conferred on him the degree of LL.D. It was the only mark of distinction he ever received from any University. Before he gave his lectures he visited Edgeworthstown, as we learn in a letter from Maria to Miss Buxton:

Davy spent a day here last week, and was as usual full of entertainment and information of various kinds. He has gone to Connemara, I believe, to fish, for he is a little mad about fishing; and very ungrateful it is of me to say so, for he sent to us from Boyle the finest trout! and a trout of Davy's catching is, I presume, worth ten trouts caught by vulgar mortals.

To his mother he writes:

Ballina, Ireland, *October 24th*.

MY DEAR MOTHER,—I am safe and well, in a remote and beautiful part of Ireland, where I have been making an excursion with two of my friends. I shall return to Dublin in two or three days, and shall be very glad to hear from you or my sisters there. I hope you are all well and happy.

I heard from John a few days ago; he was quite well and in good spirits.

The laboratory in Dublin, which has been enlarged so as to hold 550 people, will not hold half the persons who desire to attend my lectures. The 550 tickets issued for the course by the Dublin Society, at two guineas each, were all disposed of the first week; and I am told now that from ten to twenty guineas are offered for a ticket.

This is merely for your eye; it may please you to know that your son is not unpopular or useless. Every person here, from the highest to the lowest, shows me every attention and kindness.

I shall come to see you as soon as I can. I hear with infinite delight of your health, and I hope Heaven will continue to preserve and bless a mother who deserves so well of her children.

I am your very affectionate son
H. DAVY.

My kindest love to my sisters and aunts.

But Davy's affections at the moment were not wholly spent upon his kindred, and another mistress than Science had become the object of his devotion. The "little madness" of which Maria Edgeworth wrote was always a vulnerable point with Davy, for he followed the calling of the Apostles with all the zeal and ardour he gave to philosophy, and to engage him upon the subject of angling was a more direct road to his sympathies than to talk to him of science.

The wooing began in this wise:

> Mr Davy regrets that he cannot send Walton to Mrs Apreece this morning. He did not recollect that he had lent the book to a friend who lives a little way out of town. He will send honest Isaac to Mrs Apreece tomorrow or Thursday.
>
> Mrs Apreece is already of the true faith of the genuine angler, the object of whose art and contemplation is to exalt spirit above matter, to enable the mind to create its own enjoyments and to find society even in the bosom of Nature.

Matters went on apace. Shortly afterwards we read:

> I return the ticket. I begin to like the opera from association. The same association would, I think, make me love a desert, and perhaps, in a long time, might make me an admirer of routs.

Again:

> To avoid studiously what other people seek would have the semblance of affectation and though sincerely I have no ambition to shine in courts or to become a courtier; yet I have sympathy more than enough to wish to be where you like to go.

On another occasion he wrote:

> I find an invitation from Mr T—— on my return last night for Wednesday. Pray do you go to the Miss Ch——'s to-night or to Miss S——'s tomorrow night? I wish to know as you are my magnet (though you differ from a magnet in having no repulsive point) and direct my course. Your society always delightful to me is really at this moment balm to a wounded mind.

The following is a New Year's Day letter written to arrive on January 1st, 1812:

I hope the cold weather has not increased your indisposition and that the foggy sky has not made you melancholy. I trust you are now well and happy: I give myself pleasure by believing that you are.

I have a motive for writing this day besides that of doing what I like. I find that Friday the 10th is a Royal Society Club day and that I ought to dine with the Club. All other days are yours and *that* shall be yours if you command it, but I know you wish me to do what I *ought* to do, and you now cannot doubt the exclusive nature of your influence and the absolute nature of your power.

I spent the last two days very pleasantly at Wilderness, Lord Camden's; there was a very agreeable social party and a Christmas country ball: a fine park had lost its beauty from the old age of the year and everything was white; the circle round the fire had in consequence more charms and my friend and I left it this morning very well amused.

Today we celebrate the old Mr Children's birthday who is 70. He bears his years healthfully and joyfully. Such winter's days as his are rather to be desired than feared—sunny, calm and warm.

I hope, my darling friend, that you bear no uneasiness in your kind and good heart and that you give its true meaning to my unlucky sentence. Indeed I never in the whole course of our social converse ever intended to offend you or give you a moment of uneasiness and I do not think I should feel anything *long painful* that I thought would promote your happiness even though it should require from me the greatest of all sacrifices. You know what this is and I trust you will never oblige me to make it.

I go on Thursday to a wild part of Kent to shoot pheasants: the house is Mr Hodges, the post-town Cranbrook. I shall accompany Children to town on Sunday; and I hope you will permit me to see you that evening if I come in time, or Monday morning. I am going on steadily for three hours a day with Radiant Heat and Light. I might petition for one of your distant beams of light. You know it would delight me; but whether it comes or no you shall not cease to be my sun.

These letters, with many others addressed by him to the lady, are now before me. They had been carefully tied up and preserved, and are all dated by her on the back—even down to the little missives sent across from Albemarle Street to Berkeley Square, where she resided. From the number and frequency of these it is evident that the porter suffered from no lack of exercise. After her death in 1855 these letters came into the possession of Dr John Davy, together with other papers, and some have been published already in his "Fragmentary Remains." The correspondence is of especial interest from the sidelight it throws on Davy's disposition and character. Many of the letters are delightful in tone and feeling; not even Amadis de Gaul, that cream and flower of gentility, or that mirror of chivalry, the Knight of the Woful Figure, could have been more courteous in bearing, or have shown a warmer and at the same time a more deferential admiration of the lady he wooed. But the world, after all, has no concern with their tender confidences. It is sufficient to say that Davy's letters are such as might be expected from his ardent temperament and active imagination; from his love of natural scenery; his faculty of happy expression, and graphic power of description.

Early in 1812 Sir Joseph Banks, whose constant thought was of and for the Royal Society, thus wrote to his friend Sir George Stanton:

> The Royal Society has been well supplied with papers, and continues to be so. Davy, our secretary, is said to be on the point of marrying a rich and handsome widow, who has fallen in love with Science and marries him in order to obtain a footing in the Academic Groves; her name is Apreece, the daughter of Mr Carr, [Kerr] who made a fortune in India, and the niece of Dr Carr, [Kerr] of Northampton. If this takes place, it will give to science a kind of new *éclat*; we want nothing so much as the countenance of tile ladies to increase our popularity.

The lady was the widow of Shuckburgh Ashby Apreece, the eldest son of Sir Thomas Apreece; she was the daughter and heiress of Charles Kerr of Kelso, who had been secretary to Lord Rodney, and had made a fortune in the West Indies. She was also a "far-away cousin" of Sir Walter Scott, and on the occasion of his tour in the Hebrides with his family, "his dear friend and distant relation," as he calls her, accompanied them. She had been, he says, "a lioness of the first magnitude in Edinburgh" during the preceding winter; and in one of his letters to Byron in 1812, inviting him to Abbotsford, he mentions as one of the visitors that would make his house attractive "the fair or shall I say the sage Apreece that was, Lady Davy that is, who is soon to show us how much science she leads captive in Sir Humphry; so your Lordship sees, as the citizen's wife says in the farce, 'Threadneedle Street has some charms,' since they procure us such celebrated visitants." How Scott regarded her is further indicated in the letters which he addressed to her on the occasion of his son's marriage, and during the financial crash which overwhelmed him.

When the marriage was arranged Davy thus wrote to his mother:

My dear Mother,—You possibly may have heard reports of my intended marriage. Till within the last few days it was mere report. It is I trust now a settled arrangement. I am the happiest of men, in the hope of a union with a woman equally distinguished for virtues, talents and accomplishments …

You, I am sure, will sympathise in my happiness. I believe I should never have married, but for this charming woman, whose views and whose tastes coincide with my own, and who is eminently qualified to promote my best efforts and objects in life …

I am your affectionate son,
H. Davy.

In the following letter to Dr John Davy, who was then in Edinburgh as a student of medicine, we have also the announcement of another event:

Friday, *April* 10*th*, 1812.

My dear Brother,—You will have excused me for not writing to you on subjects of science. I have been absorbed by arrangements on which the happiness of my future life depends. Before you receive this these arrangements will, I trust, be settled; and, in a few weeks, I shall be able to return to my habits of study and of scientific research.

I am going to be married tomorrow; and I have a fair prospect of happiness, with the most amiable and intellectual woman I have ever known.

The Prince Regent, unsolicited by me, or by any of my intimate friends, was pleased to confer the honour of knighthood on me at the last *levée*. This distinction has not often been bestowed on scientific men; but I am proud of it, as the greatest of human geniuses bore it; and it is at least a proof that the court has not overlooked my humble efforts in the cause of science.

I have discovered pure phosphorous acid (a solid body, very volatile); and a pure hydro-phosphorous acid, containing two proportions of water and four of phosphorous acid, and decomposing by heat into phosphoric acid and a new gas containing four proportions of hydrogen and one of phosphorus …

Pray address to me Sir H. Davy, Beechwood Park, near Market St Alban's.

Believe me, my dear John, I shall always take the warmest interest in your welfare and happiness, and will do everything to promote your views. I shall have some ideas on your studies soon to communicate.

I am, my dear brother most affectionately yours
H. Davy.

He was knighted by the Prince Regent at a *levée* held at Canton House on the 8th April, 1812, being the first person on whom that honour was conferred by the Regent. On the following day he delivered his farewell lecture as Professor of Chemistry at the Royal Institution. It was on the Metals, and a report of it is contained in Faraday's manuscript notes before referred to. Faraday says:

> Having thus given the general character of the metals, Sir H. Davy proceeded to make a few observations on the connection of science with the other parts of polished and social life. Here it would be improper for me to follow him. I should merely injure and destroy the beautiful, the sublime observations that fell from his lips. He spoke in the most energetic and luminous manner of the advancement of the arts and sciences, of the connection that had always existed between them and other parts of a nation's economy. He noticed the peculiar congeries of great men in all departments of life that generally appeared together, noticed Anaximander, Anaximenes, Socrates, Newton, Bacon, Elizabeth, etc., but, by an unaccountable omission, forgot himself, though I venture to say no one else present did.
>
> During the whole of these observations his delivery was easy, his diction elegant, his tone good, and his sentiments sublime.

Two days afterwards he was married, and Lady Davy and he passed most of the spring and summer in the North of England and in Scotland, on a round of visits, cultivating those patrician instincts and susceptibilities to the charms of rank that his new station served to accentuate.

Writing to Miss Margaret Ruxton, Maria Edgeworth says:

> I suppose you have heard various *jeux d'esprit* on the marriage of Sir Humphry Davy and Mrs Apreece? I scarcely think any of them worth copying.

But she gives the following:

> Too many men have often seen
> Their talents underrated
> But Davy owns that his have been
> Duly *Apreeciated*.

Shortly after his wedding he wrote to his brother John:—

> I communicated to you in a former letter, my plans, as they were matured. I have neither given up the Institution, nor am I going to France; and, wherever I am,

I shall continue to labour in the cause of science with a zeal not diminished by increase of happiness and (with respect to the world) increased independence.

I have just finished the first part of my *Chemistry* to my own satisfaction, and I am going to publish my *Agricultural Lectures* for which I am to get 1,000 guineas for the copyright and 50 guineas for each edition, which seems a fair price …

I was appointed Professor (honorary) to the Institution, at the last meeting. I do not pledge myself to give lectures … If I lecture it will be on some new series of discoveries, should it be my fortune to make them; and I give up the *routine* of lecturing, merely that I may have more time to pursue original inquiries, and forward more the great objects of science. This has been for same time my intention, and it has been hastened by my marriage.

I shall have great pleasure in making you acquainted with Lady P. She is a noble creature (if I may he permitted so to speak of a wife), and every day adds to my contentment by the powers of her understanding, and her amiable and delightful tones of feeling.

The allusion to the Institution is thus more circumstantially dealt with in the following Minutes of the Meetings of the Managers:

May 11, 1812. Mr Hatchett reported that Sir H. Davy, though he cannot pledge himself to deliver lectures, will be willing to accept the offices of Professor of Chemistry and Director of the Laboratory and Mineralogical Collection without salary.

Following which we read:

That the Managers hear with great regret the notification which they have just received that Sir H. Davy cannot pledge himself to continue the lectures which he has been accustomed to deliver with so much honour to the Institution and advantage to the public; but at the same time, they congratulate themselves on the liberal offer which Sir Humphry Davy has made to superintend the chemical department, and to assist and advise any lecturer the Managers may be pleased to appoint.

The Managers thereupon ordered a special general meeting to nominate him Professor of Chemistry, and he was elected on June 1st. How necessary Davy was to the very existence of the Institution may be gleaned from the fact that the balance in its favour at the end of the year was £3 9s 11d.

The *Chemistry* above referred to is his *Elements of Chemical Philosophy*, which was published a few months after his marriage, with a dedication to Lady Davy. She is asked to receive it as a proof of his ardent affection, which must be unalterable, as it is founded upon the admiration of her moral and intellectual qualities. The work was begun in the autumn of 1811, and was composed with great rapidity, the "copy" being sent to the press as it left his pen. The introductory part on the History of Chemistry, and that on the General Laws of Chemical Changes and on Radiant or Ethereal Matter, and probably some other portions, are either transcripts or amplifications of his Royal Institution lectures. Other sections are avowedly based upon his own work as published in the *Philosophical Transactions*. Indeed, it was remarked by a critic that the work could never be completed upon the plan on which it was commenced, which was little less than a system of chemistry in which all the facts were to be verified by the author.

Thomas Young, his former colleague at the Royal Institution, in the *Quarterly Review* for September, 1812, thus speaks of it:

> With all its excellencies this work must be allowed to bear no inconsiderable marks of haste, and we would easily have conjectured, even if the author had not expressly told us so in his dedication, that the period employed on it has been the 'happiest of his life.' …
>
> The style and manner of this work are nearly the same with those of the author's lectures delivered in the theatre of the Royal Institution. They have been much admired by some of the most competent judges of good language and good taste, and it has been remarked that Davy was born a poet, and has only become a chemist by accident. Certainly the situation in which he was placed induced him to cultivate an ornamented and popular style of expression and embellishment, and what was encouraged by temporary motives has become natural to him from habit. Hence have arisen a multitude of sentimental reflections and appeals to the feelings, which many will think beauties and some only prettinesses; nor is it necessary for us to decide in which of the two classes of readers we wish ourselves to be arranged, conceiving that in matters so indifferent to tile immediate object of the work a great latitude may be allowed to the diversity of taste and opinion.

Despite its egoism and the obvious marks of haste and imperfection it displays, the work may still be read with interest by the chemical student. We would recommend him before perusing it to study Dalton's *New System of Chemical Philosophy*, and he will gain a vivid impression of the extraordinary

strides which the science had made during the four years which intervened
between the publication of these memorable books. Each work, too, is
strongly typical of its author, and reflects in the most striking manner the
range and limitations of his powers and the characteristics of his genius.

Towards the middle of October Davy returned to town. In a letter written
to his friend Children, from Edinburgh, he says:

> I have received a very interesting letter from Ampère. He says that a combina-
> tion of chlorine and azote has been discovered at Paris, which is a fluid, and
> explodes by the heat of the hand; the discovery of which cost an eye and a
> finger to the author. He gives no details as to the mode of combining them. I
> have tried in my little apparatus with ammonia cooled very low, and chlorine,
> but without success.

The substance here referred to is nitrogen chloride, one of the most formidable explosives known to chemists, and which seriously maimed Dulong, its
discoverer, as stated. The "little apparatus" refers to a portable chemical chest
which accompanied Davy on all his travels. Any new combination of nitrogen
was certain to attract his immediate attention. He seems to have remained to
the last convinced that nitrogen would turn out to be a non-elementary substance, and it is remarkable how eagerly he caught at any hint or surmise which
appeared likely to afford support to his conjecture. He at once repeated Dulong's
experiments in Children's laboratory at Tunbridge, and succeeded in obtaining
considerable information concerning the chemical and physical properties of
this extraordinary substance, when he was wounded in the eye by its explosion.

He thus breaks the news of his accident to Lady Davy:

> … Yesterday I began some new experiments to which a very interesting discov-
> ery and a slight accident put an end. I made use of a compound more powerful
> than gun-powder destined perhaps at some time to change the nature of war
> and influence the state of society. An explosion took place which has done me
> no other harm than that of preventing me from working this day [Sunday] and
> the effects of which will be gone tomorrow and which I should not mention at
> all, except that you may hear some foolish exaggerated account of it, for it really
> is not worth mentioning …

In reality the accident was more serious than he would have Lady Davy
believe, and the injury prevented him from resuming his work for some
time.

In a letter written about the middle of January, 1813, from Wimpole, where he was staying with Lord Hardwicke, he says:

> I have had another severe attack of inflammation in the eye, and was obliged to have the conjunctiva and cornea punctured. I suspect the cause was some little imperceptible fragment. I am just recovering, and hope I shall see as well soon as with the other eye.

In the following April he was sufficiently recovered to resume the study of Dulong's compound, and in a letter to Sir Joseph Banks, dated June 20th, 1813, and subsequently published in the *Philosophical Transactions*, he gives a number of details concerning its nature and composition. He accurately determined its specific gravity—*viz.* 1.653—but although he made a number of determinations of the amounts of its constituents by various methods, his deduction that it consisted of one proportion of nitrogen to four of chlorine was incorrect. The experiments of Gatterrmann, made with great skill and courage, have conclusively shown that the compound is, as long surmised, a trichloride of nitrogen.

At about the same period, as we learn from a letter to his brother, dated April 4th, 1813, he attacked the chemistry of fluorine:

> I am now quite recovered, and Jane [Lady Davy] is very well, and we have both enjoyed the last month in London. I have been hard at work. I have expelled fluorine from fluate of lead, fluate of silver, and fluate of soda by chlorine. It is a new acidifier, forming three powerful acids; hydro-fluoric, silicated fluoric, and fluo-boric. It has the most intense energies of combination of any known body, instantly combining with all metals, and decomposing glass. Like the fabled waters of the Styx, it cannot be preserved, not even in the ape's hoof. We have now a triad of supporters of combustion.

The results of Davy's work were communicated to the Royal Society on July 8th, 1813. In his paper he states that M. Ampère of Paris had furnished him with many ingenious and original arguments in favour of the analogy between the muriatic and fluoric compounds, based partly upon his (Davy's) views of the nature of chlorine, and partly upon reasonings drawn from the experiments of Gay Lussac and Thenard. After a short account of the main properties of the silicated fluoric acid gas (silicon fluoride), discovered by Scheele, fluoric acid (hydrofluoric acid), discovered by Scheele but first obtained pure by Gay Lussac and fluoric acid (boron fluoride), discovered by Gay Lussac and. Thenard, he states that, on the hypothesis of M. Ampère:

the silicated fluoric acid is conceived to consist of a peculiar undecompounded principle, analogous to chlorine and oxygen, united to the basis of silica, or *silicum*; the fluo-boric acid of the same principle united to boron; and the pure liquid fluoric acid as this principle united to hydrogen.

He then seeks to put the hypothesis to the test of experiment by combining fluoric acid with ammonia in a platinum apparatus; the white solid substance he obtained—so-called fluate of ammonia—contained no moisture, and hence he inferred that no water was present and that therefore fluoric acid was free from oxygen. The inference was more correct than the experiment warranted. He further found that the action of potassium upon fluate of ammonia is precisely similar to its action upon muriate of ammonia, when ammonia and hydrogen are disengaged and muriate of potassa formed. He then attempted to electrolyse solutions of hydrofluoric acid. He says:

I undertook the experiment of electrizing pure liquid fluoric acid, with considerable interest, as it seemed to offer the most probable method of ascertaining its real nature but considerable difficulties occurred in executing the process. The liquid fluoric acid immediately destroys glass, and all animal and vegetable substances; it acts on all bodies containing metallic oxides; and I know of no substances which are not rapidly dissolved or decomposed by it except [certain] metals, charcoal, phosphorus, sulphur and certain combinations of chlorine.

After various unsuccessful attempts to make tubes of sulphur and of the chlorides of lead and copper, he succeeded

in boring a piece of horn-silver in such a manner that I was able to cement a platina wire into it by means of a spirit lamp, and by inverting this in a tray of platina filled with liquid fluoric acid, I contrive to submit the fluid to the agency of electricity.

He found that the platina wire at the positive pole rapidly corroded, and became covered with a chocolate powder, and what appeared by its inflammability to be hydrogen separated at the negative pole. He tried a number of other experiments with different vessels and various electrodes, but with no better success.

He suffered great inconvenience from the fumes of hydrofluoric acid; they acted vigorously on the nails, and produced a most painful sensation when in contact with the eyes. The conclusion he drew from his experiments was that fluoric acid is "composed of hydrogen, and a substance as yet unknown in a separate form, possessed like oxygen and chlorine, of the negative electrical energy, and hence determined to the positive surface, and strongly attracted by metallic substances."

He then attempted to isolate the fluoric principle by treating various fluates in a platinum apparatus with chlorine gas, but although there was evidence of decomposition and the platinum was violently acted upon, he could obtain no new gaseous matter.

> From the general tenour of the results that I have stated, it appears reasonable to conclude that there exists in the fluoric compounds a peculiar substance, possessed of strong attractions for metallic bodies and hydrogen, and which combined with certain inflammable bodies forms peculiar acids, and which in consequence of its strong affinities and high decomposing agencies, it will be very difficult to examine in a pure form, and for the sake of avoiding circumlocution, it may be denominated fluorine, a name suggested to me by M. Ampère.
>
> It is easy to perceive in following the above theory, that all the ideas current in chemical authors respecting the fluoric combinations, must be changed. Fluor-spar, and other analogous substances, for instance, must be regarded as binary compounds of metals and fluorine.

Davy's views are now part of current chemical doctrine, and his previsions as to the nature of fluorine and its extraordinary chemical activity have been verified in the most striking manner by the admirable investigations of Moissan.

IX

DAVY AND FARADAY—IODINE

THE YEAR 1813 IS MEMORABLE in the history of the Royal Institution, from the fact that Faraday's long and honourable association with it dates from that time. The circumstances which led to this connection were subsequently stated by himself in the following letter to Dr Paris:

Royal Institution, *Dec.* 23rd, 1829.

My DEAR SIR,—You ask me to give you an account of my first introduction to Sir H. Davy, which I am very happy to do, as I think the circumstances will bear testimony to his goodness of heart.

When I was a bookseller's apprentice, I was very fond of experiment and very averse to trade. It happened that a gentleman, a member of the Royal Institution, took me to hear some of Sir H. Davy's last lectures in Albemarle Street. I took notes, and afterwards wrote them out more fairly in a quarto volume.

My desire to escape from trade, which I thought vicious and selfish, and to enter into the service of Science which I imagined made its pursuers amiable and liberal, induced me at last to take the hold and simple step of writing to Sir H. Davy, expressing my wishes, and a hope that, if an opportunity came in his way, he would favour my views; at the same time I sent the notes I had taken at his lectures.

The answer, which makes all the point of my communication, I send you in the original, requesting you to take great care of it, and to let me have it back, for you may imagine how much I value.

You will observe that this took place at the end of the year 1812, and early in 1813 he requested to see me, and told me of the situation of assistant in the laboratory of the Royal Institution, then just vacant.

At the same time that he thus gratified my desires as to scientific employment, he still advised me not to give up the prospects I had before me, telling me that Science was a harsh mistress; and, in a pecuniary point of view, but poorly rewarding those who devoted themselves to her service. He smiled at my notion of the superior moral feelings of philosophic men, and said he would leave me to the experience of a few years to set me right on that matter.

Finally, through his good efforts I went to the Royal Institution early in March of 1813, as assistant in the laboratory; and in October of the same year went with him abroad as his assistant in experiments and in writing. I returned with him in April 1815, resumed my station in the Royal Institution, and have, as you know, ever since remained there.

I am, dear Sir, very truly yours
M. FARADAY.

The answer which Faraday characteristically says makes all the point of the foregoing communication is as follows:

December 24th, 1812.

SIR,—I am far from displeased with the proof you have given me of your confidence, and which displays great zeal, power of memory, and attention. I am obliged to go out of town, and shall not be settled in town till the end of January: I will then see you at any time you wish.

It would gratify me to be of any service to yon. I wish it may be in my power.

I am, Sir, your obedient humble servant,
H. DAVY

The immediate cause of the connection was very trivial and commonplace. Mr W. Payne, whose name may be recalled in connection with Davy's memorandum respecting the state in which the Laboratory of the Institution was kept, in the latter part of February, 1813, had a disagreement with Mr Newman; the instrument-maker, and so far forgot himself as to strike that gentleman. Whereupon the Managers immediately resolved that Mr Payne should be dismissed from the Royal Institution, and that a gratuity of £10 should be paid him in consideration of his long services. Davy appears then to have called to mind the modest, bright-eyed, active youth with the pleasant smile, who had expressed his desire to devote himself to science.

In the minutes of the meeting of Managers on March 1st, 181.3, we read:

Sir Humphry Davy has the honour to inform the Managers that he has found a person who is desirous to occupy the situation in the Institution lately filled by William Payne. His name is Michael Faraday. He is a youth of twenty-two years of age. As far as Sir H. Davy has been able to observe or ascertain, he appears well fitted for the situation. His habits seem good, his disposition active and cheerful, and his manner intelligent. He is willing to engage himself on the same terms as those given to Mr Payne at the time of quitting the Institution.

Resolved—That Michael Faraday be engaged to fill the situation lately occupied by Mr Payne on the same terms.

In the minutes of the general monthly meeting of the members on April 5th, 1813, for putting in nomination from the chair the professors for the year ensuing, we read:

Sir H. Davy rose, and begged leave to resign his situation of Professor of Chemistry; but he by no means wished to give up his connection with the Royal Institution, as he should ever be happy to communicate his researches in the first instance to the Institution … and to do all in his power to promote the interest and success of this Institution. Sir H. Davy having retired, Earl Spencer moved, that the thanks of this Meeting be returned to Sir H. Davy for the estimable services rendered by him to the Royal Institution. This motion was seconded by the Earl of Darnley, and, on being put, was carried unanimously. Earl Spencer further moved, that in order more strongly to mark the high sense entertained by this Meeting of the merits of Sir H. Davy, he be elected Honorary Professor of Chemistry; which, on being seconded by the Earl of Darnley, met with unanimous approbation.

Mr Brande was subsequently elected Professor.

During the autumn Davy obtained permission from Napoleon to pass through France in the course of an extended tour on the Continent which Lady Davy and he now projected. He thus announced his intention to his mother:

Andover, *Oct.* 14, 1813.

My dear Mother,—We are just going to the Continent upon a journey of scientific inquiry which I hope will be pleasant to us and useful to the world. We go rapidly through France to Italy, and from there to Sicily; and we shall return through Germany. We have every assurance from the governments of the countries through which we pass, that we shall not be molested, but assisted. We shall stay probably a year or two ...

As soon as I have settled a plan of correspondence abroad, I will write to you, and shall hear of you from John as often as possible. As I am permitted to pass through an enemy's country, there must be no politics in any letters to me; and you had better not write except through the channel I shall hereafter point out ...

When I return I shall peacefully fix my abode for life in my own country. Pray take care of Betsy. When the wind is cold she should not think of going out. Tell Grace not to be afraid, though I am going through France. My love to Kitty, and to Grace and Betsy. I am, my dear mother, wishing you all health and happiness, your very affectionate son

H. Davy.

On October 4th we find that he reported to the Managers that

Michael Faraday had expressed a wish to accompany him on his scientific travels, but that he would not engage Mr Faraday if the Professor of Chemistry considered his services as at all essential to the Institution, or if the Managers had the slightest objection to the measure.

Mr Brande reported that arrangements could be made to allow Mr Faraday to leave,

and that as he had shown considerable diligence and attention in cleaning and arranging the mineral collection he recommended his services to the Managers' attention, as this was not his immediate duty.

A few days afterwards the party, consisting of Sir H. and Lady Davy, Mr Faraday, and Lady Davy's maid, together with the chemical cabinet, crossed in a cartel from Plymouth to Morlaix. Here they were arrested, but after a week's detention, allowed to depart for Paris, where they arrived on October 27th. Nothing could exceed the cordiality and warmth of Davy's reception by the French *savants*. On November 2nd he attended a sitting of the First Class of the Institute, and was placed on the right hand of the President, who announced to the meeting that it was honoured by the presence of "Le Chevalier Davy." Each day saw some reception or entertainment in his honour. On November 10th he dined with Rumford at Auteuil. How much had happened in the ten years since last they met, and how different their situations now! Davy at the very summit of his scientific eminence, courted and caressed by society, honoured and admired by his intellectual peers; Rumford, his former patron, a broken-hearted, disappointed man about to sink into the grave, worried to death, in fact, by his wife, and the victim of the spiteful persecutions she instigated. Of the remarkable men of science whom Davy met on these occasions be has left us some slight sketches composed during his last illness, some of which are of interest to the student who desires to know something of the men whose names are as household words in the history of chemistry. Guyton de Morveau—who played such a leading part in the political Revolution of France, as well as in the revolution of its chemistry, and who, with Fourcroy, popularised the doctrines of Lavoisier whilst bringing his head to the scaffold—was found to be a gentleman of mild and conciliatory manners.

Vauquelin gave him the idea of the French chemists of another age, belonging rather to the pharmaceutical laboratory than to the philosophical one.

Nothing could be more singular than his manners, his life, and his *ménage*. Two old maiden ladies, Mdlles. de Fourcroy, sisters of the professor of that name, kept his house. I remember the first time that I entered it, I was ushered into a sort of bed-chamber, which likewise served as a drawing-room. One of these ladies was in bed, but employed in preparations for the kitchen; and was actually paring truffles ... Nothing could be more extraordinary than the simplicity of his conversation;—he had not the slightest tact, and even in the presence of young ladies, talked of subjects which, since the paradisaical times, never have been the objects of common conversation.

CUVIER had even in his address and manner the character of a superior man;—much general power and eloquence in conversation, and a great variety of information on scientific as well as popular subjects. I should say of him, that he is the most distinguished man of *talents* I have known; but I doubt if he is entitled to the appellation of a man of genius.

HUMBOLDT was one of the most agreeable men I have ever known, social, modest, full of intelligence, with facilities of every kind: almost *too fluent* in conversation. His travels display a spirit of enterprise. His works are monuments of the variety of his knowledge and resources.

Of his great rival his comment is as follow:

GAY LUSSAC was quick, lively, ingenious, and profound, with great activity of mind and great facility of manipulation. I should place him at the head of living chemists of France.

BERTHOLLET was a most amiable man; when the friend of Napoleon even, always good, conciliatory and modest, frank and candid. He had no airs, and many graces. In every way below La Place in intellectual powers, he appeared superior to him in moral qualities. Berthollet had no appearance of a man of genius; but one could not look on La Place's physiognomy without being convinced that he was a very extraordinary man.

All accounts appear to show that Davy hardly treated his hosts with the cordiality and respect they extended to him. His Chauvinism seemed to get the better of his courtesy. There was, it is said, a flippancy in his manner and a superciliousness and hauteur in his deportment which surprised as much as they offended. Napoleon, with characteristic bluntness, told one of the members of the Institute that he had heard the young English chemist had a poor opinion of them all. Dr Paris, who could certainly speak from personal knowledge, states that Davy's unfortunate manner was not so much the expression of a haughty consciousness of superiority as the desire to conceal a *mauvaise honte and gaucherie*— an ungraceful timidity he could never conquer, and which often led him to force himself into a state of effrontery and with a violence of effort which passed for a sally of pride or the ebullition of temper.

Whatever Davy's manner might have been, it was not allowed to affect the admiration felt for his genius, and on December 13th, 1813, he was with practical unanimity elected a Corresponding Member of the First Class of the Institute.

During the last week of the preceding November Ampère had given Davy a small quantity of a substance which he had obtained from Clement, and which had been discovered by Courtois, a soap-boiler and manufacturer of saltpetre in Paris, in kelp or the ashes of sea-weeds. The substance had the extraordinary property of giving a violet-coloured vapour, but its true nature and relations were unknown, and it was commonly designated as X. Although actually known for some time previously, the first public notice of its existence was made by Clement at a meeting of the Institute on November 29th, 1813, and at the meeting on December 6th Gay Lussac presented a short note on the substance, to which he gave the name *iode*, and stated that it had analogies to chlorine. A week later—that is, on the day of Davy's election to the Institute—a letter from him to Cuvier was read, in which he gave a general view of the chemical characters of the body; and on January 20th, 1814, a paper by him, dated Paris, December 10th, 1813, and entitled "Some Experiments and Observations on a new Substance which becomes a violet-coloured Gas by Heat," was read to the Royal Society.

After reciting the above facts he explains why he has ventured to take up a subject on which Gay Lussac was still engaged. The explanation was no doubt necessary; he had evidently not forgotten Gay Lussac's intrusion into his own field of work on the occasion of the discovery of the metals of the alkalis. He first draws attention to the peculiarities of the combination of the new substance with silver; this, he shows, is markedly different from silver chloride. He then forms this compound synthetically; forms also the combination with potassium by direct union, and describes its properties; studies the action of chlorine on the new substance, and notes the formation of the yellow solid chloride and the mode of its decomposition by water; prepares a number of metallic compounds; studies the action of the new substance on phosphorus, the nature of the product, and its mode of decomposition by water, with formation of the white crystalline phosphonium iodide and hydriodic acid gas. By acting on this gas with potassium he shows that it yields half its volume of hydrogen and forms the same product as by the direct union of the alkali metal with the new substance. He further finds that this gas is formed when the new substance and hydrogen are passed through a heated tube; it has a very strong attraction for water, which dissolves it to a large extent, and the concentrated solution rapidly becomes tawny. When the new substance is treated with potash solution it forms the same product as by its direct union with potassium, together with a salt precisely similar to potassium hyperoxymuriate, and which, like that salt, is decomposed when heated, with evolution of oxygen. He shows that the new substance is expelled from its compounds when these are heated with chlorine. He studies the nature of the black fulminating compound discovered by Desormes and Clement by acting on the new substance with solution of ammonia, and concludes that it is analogous to the detonating oil of Dulong. He attempts to determine the combining proportion of the new substance, on the assumption that its compounds are analogous to those of chlorine, but he has to admit that his experiments have been made upon quantities too small to afford exact results. Nevertheless they prove that the value is much higher than those of the simple inflammable bodies, and higher even than those of most of the metals. He further shows that the combination with hydrogen must be one of the heaviest elastic fluids existing.

From all the facts that have been stated, there is every reason to consider this new substance as *an undecompounded body*. In its specific gravity, lustre, colour, and the high number in which it enters into combinations it resembles the metals; but in all its chemical agencies it is more analogous to oxygen and chlorine; it is a non-conductor of electricity, and possesses, like these bodies, the negative electrical energy with respect to metals, inflammable and alkaline substances, and hence when combined with these substances in aqueous solution and electrized in the voltaic circuit, it separates at the positive surface; but it has a positive energy with respect to chlorine … It agrees with chlorine and fluorine in forming acids with hydrogen.

The name *ione* has been proposed in France for this new substance from its colour in the gaseous state, from ἴον, viola; and its combination with hydrogen has been named *hydroionic acid*. The name *ione*, in English, would lead to confusion, for its compounds would be called *ionic* and *ionian*. By terming it iodine, from ἰώδης, violaceous, this confusion will be avoided, and the name will be more analogous to chlorine and fluorine.

The rapidity with which Davy ascertained the properties and relations of the new substance was characteristic of him. A fortnight's work—done partly at his hotel and partly in the laboratory of the young Chevreul, amidst a succession of interruptions caused by *fêtes*, *levées*, and visits of ceremony— sufficed to accumulate the material for his Royal Society paper, in which he gives with unerring precision, in spite of the small quantity of the matter at his disposal, the broad outlines of the chemistry of iodine. The paper shows him at his best: he seems to have seized, as if by instinct, upon the central fact of the analogy of iodine to chlorine, and he worked out the clue with a perspicacity and insight worthy of his genius.

As may be surmised, Davy's action hardly contributed to his popularity with a certain section of the *savants* of Paris. Gay Lussac and Thenard were extremely angry with Ampère and Clement for having given him the material for his investigation, and the feeling broke out after the publication of Gay Lussac's memoir in the *Annales de Chimie* in 1814. Davy in a note published in the Journal of the Royal Institution says:

Who had most share in developing the chemical history of that body [iodine] must be determined by a review of the papers that have been published upon it, and by an examination of their respective dates. When M. Clement showed Iodine to me, he believed that the hydriodic acid was muriatic acid; and M. Gay Lussac, after his early experiments, made originally with M. Clement, formed the same opinion, and *maintained* it, when I *first* stated to him my belief that it was a new and peculiar acid, and that Iodine was a substance analogous in its chemical relations to Chlorine.

Davy left Paris towards the end of December, passing into Auvergne and thence to Montpellier, where he resumed his work on iodine. He then went to Genoa, where he made some experiments on the electricity of the torpedo, and about the middle of March arrived at Florence. In a letter to his brother John he says:

I have worked a good deal on iodine and a little on the torpedo. Iodine had been in embryo for two years. I came to Paris; Clement requested me to examine it, and he believed that it was a compound, affording muriatic acid. I worked upon it for some time, and determined that it was a new body, and that it afforded a peculiar acid by combining with hydrogen, and this I mentioned to Gay Lussac, Ampere, and other chemists. The first immediately 'took the word of the Lord out of the mouth of His servant,' and treated this subject as he had treated potassium and boron. The paper which I sent to the Royal Society on iodine I wrote with Clement's approbation and a note published in the 'Journal de Physique' will vindicate my priority. I have just got ready for the Royal Society a second paper on this fourth supporter of combustion.

The old theory is nearly abandoned in France. Berthollet, with much candour, has decided in favour of chlorine. I know no chemist but Thenard who upholds it at Paris, and he upholds it feebly, and by this time, probably, has renounced it.

I doubt if the organ of the torpedo is analogous to the pile of Volta. I have not been able to gain any chemical effects by the shock sent through water; but I tried on small and not very active animals. I shall resume the inquiry at Naples, where I hope to be about the middle of May. In my journey I met with no difficulties of any kind, and received every attention from the scientific men of Paris, and the most liberal permission to go where I pleased front the government.

I lived very much with Berthollet, Cuvier, Chaptal, Vauquelin, Humboldt, Morveau, Clement, Chevreul, and Gay Lussac. They were all kind and attentive to me; and, except for Gay Lussac's last turn of publishing without acknowledgement what he had first learnt from me, I should have had nothing to complain of; but who can control self-love?

It ought not to interfere with truth and justice; but I will not moralise nor complain. Iodine is as useful an ally to me as I could have found at home.

At Florence he worked in the laboratory of the Accademia del Cimento on iodine and on the diamond. The results of his work on iodine he embodied in a paper read to the Royal Society on June 16th, 1814, which deals mainly with the iodates, or, as he preferred to call them, the *oxyiodes*. The object of his work on the diamond was to determine whether any peculiar matter separated from it during its combustion, and whether the gas formed in the process was precisely the same in its chemical nature as that produced by the combustion of plumbago and charcoal. At Florence the use of the great burning-glass originally employed in the trials on the action of solar heat on the diamond instituted by Cosmo III, Grand Duke of Tuscany; he completed the research in the laboratory of the Accademia dei Lincei at Rome.

From the results of his different experiments, which were communicated to the Royal Society on June 23rd, 1814, it appeared that the diamond affords no other substance by its combustion in oxygen than pure carbonic acid gas, and that the only chemical difference perceptible between diamond and the purest charcoal is that the latter contains a minute proportion of hydrogen. "But," he asks, "can a quantity of an element, less in some cases than 1/50000 part of the weight of the substance, occasion so great a difference in physical and, chemical characters?" This he concludes is most unlikely, for, as he points out, even when the minute quantity of hydrogen is expelled by heating the charcoal in chlorine, the specific differences remain.

The doctrine at that time current, and which seemed indeed almost axiomatic, "That bodies cannot be exactly the same in composition or chemical nature, and yet totally different in all their physical properties," received its first great shock. Davy's work, no doubt, paved the way for the, recognition of the fact of allotropy, and thereafter of isomerism.

In May he went to Naples and made his first ascent of Vesuvius, which he revisited on several subsequent journeys. He commissioned one of the guides to inform him from time to time of the condition of the volcano, and the man's letters, in spite of their phonetic address—"Siromfredevi-Londra"— duly found their way to Albemarle Street. He also interested himself in the excavations at Pompeii instituted by direction of Murat, then King of Naples, and he performed a number of experiments on the colours used by the ancients in painting, an account of which was communicated to the Royal Society on February 23rd, 1815.

He then passed northwards with the intention of spending the summer at Geneva. On his way he called at Milan to pay his respects to Volta. Of this visit he wrote:

Volta I saw at Milan, in 1814, at that time advanced in years,—I think nearly seventy and in bad health. His conversation was not brilliant; his views rather limited, but marking great ingenuity. His manners were perfectly simple. He had not the air of a courtier, or even of a man who had seen the world.

If Dr Paris's story is to be credited, the lack of brilliancy in the conversation of the great Italian physicist may be attributed to the circumstances of this meeting. Davy, we are told, had written to announce his intended visit, and on the appointed day and hour Volta, in full dress, awaited his arrival.

On the entrance of the great English philosopher into the apartment, not only in *déshabille*, but in a dress of which an English artisan would have been ashamed, Volta started back in astonishment, and such was the effect of his surprise, that he was for some time unable to address him.

The party remained at Geneva until the middle of September, partaking freely of the intellectual life which that charming city afforded. Here he met Saussure Pictet, De la Rive, Madame de Staël, Benjamin Constant, Necker, and Talma, whose society he greatly enjoyed. With the approach of winter he returned to Italy *via* the Brenner and Venice, and on November 2nd arrived at Rome, where he remained until March 1st, 1815, occupying himself with his inquiry into the composition of ancient colours. In this he was greatly assisted by the kindness of his friend Canova, the celebrated sculptor, who was then charged with the care of the works connected with ancient art in Rome, and who supplied him with material from the colours found in the Baths of Titus and of Livia and other palaces and baths of ancient Rome and Pompeii. Davy's memoir, which appears in the *Philosophical Transactions* for 1815, displays considerable antiquarian and bibliographical research, and, considering his limited means, much analytical skill and ingenuity. The ancient reds he found to consist of minium, several varieties of iron ochre, and vermilion or cinnabar. The yellows were mixtures of ochres and chalks, or of ochre with minium. He was unable to discover that orpiment was used; a deep orange yellow on stucco in the ruins near the monument of Caius Cestius consisted of a mixture of massicot and minium. The blues were mainly mixtures of the Egyptian or Alexandrine blue, with more or less chalk. This Egyptian blue, he found, was a *frit,* made by heating soda, sand, and copper, either used as an ore or as metal. He gives a method of making it, and speaks highly of its permanence and beauty. The greens were, as a rule, compounds of copper. The exact nature of the purples he was unable to determine; they were prob-

ably organic, but whether obtained from shell-fish or madder could not be
ascertained. The purplish reds in the Baths of Titus were found to be mixtures
of red ochres, and the blues were copper compounds. The blacks and browns
were mixtures of carbonaceous matter with oxides of iron, or manganese. The
whites were mainly chalk, or occasionally clay; cerusse, or white-lead, was
apparently not used.

Before leaving Italy he again went to Naples, for the purpose of witnessing
Vesuvius in eruption, and on several occasions he was as near the crater as he
could get. He left Naples on March 21st, and came home by way of Verona,
Innsbruck, Ulm, Stuttgart, Heidelberg, and the Rhine, arriving in London
April 23rd, 1815. A few days after his arrival he wrote to his mother:

> We have had a very agreeable and instructive journey and Lady Davy agrees
> with me in thinking that England is the only country to *live* in, however inter-
> esting it may be to *see* other countries.
>
> I yesterday bought a good house in Grosvenor Street, and we shall sit down
> in this happy land.
>
> I beg you to give my best and kindest love to my sisters, and to remember me
> with all affection to my aunts.

Faraday was again engaged as assistant in the laboratory of the Royal
Institution and superintendent of the apparatus (at a salary of 30s. a week),
and was accommodated with apartments at the top of the house.

In Dr Bence Jones's *Life of Faraday* we have more detailed information
concerning this tour, derived from the journal which Faraday kept whilst
he was abroad. Faraday describes in considerable detail the life in Paris
and the work on Iodine; we have accounts of Chevreul's laboratory at the
Jardin des Plantes, and of Gay Lussac's lectures at the École Polytechnique;
of the work on the torpedo at Genoa; of the combustion of the diamond
at the Accademia del Cimento, and a description of the great burning-
glass, and how it was actually employed; of the experiments of Morichini
on the alleged magnetisation of a needle by the solar rays; of his meeting
Volta—"an hale, elderly man, bearing the red ribbon, and very free in conversation"; of the work at Rome on chlorous oxide and iodic acid, and on
the pigments employed by the ancients.

The constant presence of Sir Humphry Davy," wrote Faraday to his friend
Abbott, "is a mine inexhaustible of knowledge and improvement." But he
adds: "I have several times been more than half decided to return hastily

home; but second thoughts have still induced me to try what the future may produce … the glorious opportunities I enjoy of improving in the knowledge of chemistry and the sciences continually determine me to finish this voyage with Sir H. D. But if I wish to enjoy these advantages I have to sacrifice much, and though these sacrifices are such as an humble man would not feel, yet I cannot quietly make them.

Faraday's troubles arose from his anomalous position in the party. When Davy elected to go abroad, he arranged to take his valet with him; but at the eleventh hour this man, moved by the tears of his wife—to whom the "Corsican Ogre" was a kind of bogey—refused to proceed. "When Sir H. informed me of this circumstance," says Faraday, "he expressed his sorrow at it, and said—that if I would put up with a few things on the road until he got to Paris, doing those things which could not be trusted to strangers or waiters … he would get a servant … At Paris he could find no servant to suit him," nor was he more successful at Montpellier or at Genoa. It was, doubtless, difficult at this period to find a man in such places who understood English and was in other respects suitable. Faraday goes on to say:

> Sir Humphry has at all times endeavoured to keep me from the performance of those things which did not form a part of my duty, and which might be disagreeable … I should have but little to complain of, were I travelling with Sir Humphry alone, or were Lady Davy like him; but her temper makes it oftentimes go wrong, with me, with herself and with Sir H …
>
> She likes to show her authority, and at first I found her extremely earnest in mortifying me. This occasioned quarrels between us, at each of which I gained ground and she lost it; for the frequency made me care nothing about them, and weakened her authority, and after each she behaved in a milder manner.

How Davy and his wife appeared to the world at this time may be seen from the following extracts from Ticknor's Life:

> 1815. *June* 13.—I breakfasted this morning with Sir H. Davy, of whom we have heard so much in America. He is now about thirty-three [he was actually thirty-seven], but with all the freshness and bloom of twenty-five, and one of the handsomest men I have seen in England. He has a great deal of vivacity—talks rapidly, though with great precision—and is so much interested in conversation that his excitement amounts to nervous impatience, and keeps him in constant motion. He has just returned from Italy, and delights to talk of

it; thinks it, next to England, the finest country in the world, and the society of Rome surpassed only by that of London, and says he should not die contented without going there again.

15 *June*.—As her husband had invited me to do, I called this morning on Lady Davy. I found her in her parlour, working on a dress, the contents of her basket strewed about the table, and looking more like home than anything since I left it. She is small, with black eyes and hair and a very pleasant face, an uncommonly sweet smile; and when she speaks has much spirit and expression in her countenance. Her conversation is agreeable, particularly in the choice and variety of her phraseology, and has more the air of eloquence than I have ever heard before from a lady. But, then, it has something of the appearance of formality and display, which injures conversation. Her manner is gracious and elegant; and though I should not think of comparing her to Corinne yet I think she has uncommon powers.

In Henry Crabb Robinson's Diary we read, under date May 31st, 1813:

Dined with Wordsworth at Mr Carr's. Sir Humphry and Lady Davy there. She and Sir H. seem to have hardly finished their honeymoon. Miss Joanna Baillie said to Wordsworth, 'We have witnessed a picturesque happiness.'

In 1815 it was very evident the honeymoon had waned and that the picturesque happiness was at an end. However fitted her ladyship might be to shine in salons, at routs and fashionable gatherings, she lacked the homelier, kindlier charms which grace the *placens uxor*. An accomplished woman, of fastidious taste, fond of study, upright in her dealings, and charitable to the poor, she was withal cold and unsympathetic, self-willed and independent, "fitted to excite admiration rather than love, and neither by nature happy in herself, or qualified to impart, in the best sense of the term, happiness to others." Such is the character given of her by Dr Davy; and he adds, "There was an oversight, if not a delusion, as to the fitness of their union" and "it might have been better for both if they had never met." It was, no doubt, from the fullness of his own experience that Davy once wrote to a friend:

Upon points of affection it is only for the parties themselves to form just opinions of what is really necessary to ensure the felicity of the marriage state. Riches appear to me not at all necessary, but competence, I think is; and after this more depends upon the temper of the individual than upon personal, or even intellectual circumstances. The finest spirits, the most exquisite wines, the nectars and ambrosias of modern tables, will be all spoilt by a few drops of bitter extract; and a bad temper has the same effect in life, which is made up, not of great sacrifices or duties, but of little things, in which smiles and kindness, and small obligations given habitually, are what win and preserve the heart; and secure comfort.

X

THE SAFETY LAMP

SHORTLY AFTER DAVY'S RETURN TO England his sympathy was enlisted in a cause which enabled him to display all the attributes of his genius, and to achieve a triumph which, while greatly enhancing his popular reputation, added no little to his scientific fame. To show him how he might be useful, was at all times a certain method of securing his interest; for, like Lavoisier, he was even more the friend of humanity than of science, and to make science serviceable to humanity was, he considered, the highest object of his calling.

During the early years of this century the country was repeatedly shocked by the occurrence of a succession of disastrous colliery explosions, especially in the north of England, attended by great destruction of life and property and widespread misery and destitution. The development of our iron-trade, the improvements in the steam-engine, and the more general application of machinery to industry had greatly stimulated the opening out of our coal-fields; and the working of coal was being extended with a rapidity that greatly aggravated the evils and dangers at all times inseparable from it. In the early days of coal-getting, when the pits were shallow and the workings compara-tively near the shafts, fire-damp, although not unheard of, was little dreaded, and explosions were rare—so rare, indeed, that when they occurred they were thought worthy of mention in the *Philosophical Transactions* of the Royal Society. As the pits became deeper, and the ways more extended, explosions became more frequent, and, at times it was impossible to work the coal, owing to the accumulation of fire-damp and its liability to "fire" at the candles of the miners. In 1732 attempts were first made to ventilate the pits by "fire-lamps" or furnaces, and by mechanical means, so as to sweep out the "sulphur" by means of fresh air. Carlisle Spedding, a little later, invented the steel mill—a contrivance by which a disc of steel was caused to revolve against a piece of flint, so as to throw off a shower of sparks sufficiently luminous to enable the miner to carry on his business.

In spite of the "spark-emitting wheel," and of the systems of ventilation introduced by Ryan, James Spedding, John Buddle, and others, "the swart demon of the mine" grew more and more formidable, and demanded a greater number of victims every year. Mechanical science would appear to have spent itself, and the mining world was gradually coming to look upon fire-damp with the fatalism with which ignorant and superstitious people regard the plague. Some of the great coal owners—powerless to do more, but afraid of the rising tide of public opinion—used their influence with the newspapers to suppress all allusion to these calamities. But many persons, especially the physicians and clergymen in the mining districts, who were witnesses of the suffering and distress which the "firing" of a mine occasioned, kept public attention alive by means of pamphlets and letters and notices to such journals as would insert their communications. One colliery—the Brandling Main or Felling Colliery, near Gateshead-on-Tyne— acquired an unenviable notoriety from the frequency with which it fired. On May 25th, 1812, an explosion occurred which killed ninety-two men and boys. No calamity of such magnitude had ever happened before in a coal mine. Eighteen months afterwards a second explosion took place by which twenty-three lives were lost. In the following year explosions occurred at Percy Main, Hebburn, and Seafield. In June, 1815, Newbottle Colliery exploded with the loss of fifty-seven men and boys, and this was immediately followed by a similar disaster at Sheriff Hill. The Rev. Mr Hodgson—the historian of Northumberland—in whose parish the Brandling Main was situated, published a particular account of the first Felling Colliery Explosion. This was widely circulated, and ultimately found its way into Thomson's *Annals of Philosophy*, which continued to print accounts of similar accidents as they occurred. At length Mr J. J. Wilkinson, a barrister resident in the Temple, suggested the formation of a society to investigate the whole subject and to seek for remedies. The Bishop of Durham and the Rev. Dr Gray, afterwards Bishop of Bristol, but then Rector of Bishopwearmouth, led the way, and ultimately the society was instituted on October 1st, 1813, with Sir Ralph Millbanke, afterwards Sir Ralph Noel as President. Its first report contains a letter from Mr John Buddle, the great authority on the ventilation of coal mines, in which he expresses his conviction that mechanical agencies are practically powerless to prevent explosions in mines subjected to sudden bursts of fire-damp, and he concludes:

> that the hopes of this society ever seeing its most desirable object accomplished must rest upon the event of some method being discovered, of producing such a chemical change upon carburetted hydrogen gas as to render it innoxious as fast as it is discharged, or as it approaches the neighbourhood of lights. In this view of the subject, it is to scientific men only that we must look up for assistance in providing a cheap and effectual remedy.

The society received a number of suggestions, for the most part wholly impracticable, and generally of the character of that of Dr Trotter, who proposed to flood the mines with chlorine. A variety of air-tight or insulated lamps were suggested by Clanny, Brandling, Murray, and others, much on the same lines as that devised by Humboldt, but none of them appears to have been seriously tried.

Under these circumstances it was decided to ask for the co-operation of Davy, and with that object Mr Wilkinson called upon him at the Royal Institution, in the autumn of 1813, but found he had left for Paris. A few months after his return the Rev. Dr Gray wrote to him on the subject, and received the following letter in reply:

> *August* 3, 1815.
> … it will give me great satisfaction if my chemical knowledge can be of any use in an enquiry so interesting to humanity, and I beg you will assure the committee of my readiness to co-operate with them in any experiments or investigations on the subject.
>
> If you think my visiting the mines can be of any use, I will cheerfully do so.
>
> I shall be here ten days longer, and on my return South, will visit any place you will be kind enough to point out to me, where I may be able to acquire information on the subject of coal gas.

Dr Gray, in reply, referred him to Mr John Buddle, of the Wallsend Colliery.

On August 24th, 1815, Mr Buddle wrote to Dr Gray:

> Permit me to offer my best acknowledgments for the opportunity which your attention to the cause of humanity has afforded me of being introduced to Sir Humphry Davy.

I was this morning favoured with a call from him, and he was accompanied by the Rev. Mr Hodgson. He made particular enquiries into the nature of the danger arising from the discharge of the inflammable gas in our mines. I shall supply him with a quantity of the gas to analyze; and he has given me reason to expect that a substitute may be found for the steel mill, which will not fire the gas. He seems also to think it possible to generate a gas, at a moderate expense, which, by mixing with the atmospheric current, will so far neutralise the inflammable air, as to prevent it firing at the candles of the workmen.

If he should be so fortunate as to succeed in either the one or the other of these points, he will render the most essential benefit to the mining interest of this country, and to the cause of humanity in particular.

After spending a few days in the district with Mr Hodgson and Dr Gray, in the course of which he saw and experimented with Dr Clanny's lamp, he went on a round of visits in Durham and Yorkshire, and arrived in London at the end of September. Early in October a quantity of fire-damp was sent to him by Mr Hodgson, the receipt of which he acknowledged on the 15th, saying:

My experiments are going on successfully and I hope in a few days to send you an account of them; I am going to be fortunate far beyond my expectations.

Four days afterwards he again wrote to Mr Hodgson stating that he had discovered

that explosive mixtures of mine-damp will not pass through small apertures or tubes; and that if a lamp or lanthorn be made air-tight on the sides, and furnished with apertures to admit the air, it will not communicate flame to the outward atmosphere.

On the 25th October he gave an account of his work to the Chemical Club. On October 30th he wrote to Dr Gray and to Mr Hodgson, giving a description of three forms of *safe lamps*. His letter to Dr Gray was as follows:

As it was the consequence of your invitation that I endeavoured to investigate the nature of the fire-damp, I owe to you the first notice of the progress of my experiments.

My results have been successful far beyond my expectations. I shall enclose a little sketch of my views on the subject; and I hope in a few days to be able to send a paper with the apparatus for the committee. I trust the *safe lamp* will answer all the objects of the collier.

I consider this at present as a private communication I wish you to examine the lamps I have had constructed, before you give any account of my labours to the committee.

I have never received so much pleasure from the result of any of my chemical labours; for I trust the cause of humanity will gain something by it.

Mr Hodgson's letter was shown to several persons, and appears to have been copied by some, on or about November 2nd, and an extract from it appeared in Dunn's *View of the Coal Trade*.

On November 9th Davy read his first paper on the subject before the Royal Society; it was entitled "On the fire-damp of coal mines, and on the methods of lighting the mines so as to prevent its explosion." After describing the manner in which his attention had been specially called to the subject, he states that he first made experiments with a variety of phosphori (Kunckel's, Canton's, and Baldwin's), and also with the electrical light in close vessels, in the hope that they might be found to afford the requisite amount of illumination; but the results were not encouraging.

After an account of the chemical characters of the fire-damp sent to him by Mr Hodgson, he describes the results of experiments on its combustibility and explosive nature, and on the degree of heat required to explode it when mixed with air. In respect of its combustibility fire-damp was found to differ most materially from the other common inflammable gases in that it required a far higher temperature to effect its inflammation or explosion. Moreover, it was found that the flame formed by the union of air and fire-damp would not pass through tubes of a certain minimum diameter;

and in comparing the power of tubes of metal and those of glass, it appeared that the flame passed more readily through glass tubes of the same diameter; and that explosions were stopped by metallic tubes of one-fifth of an inch when they were an inch and a half long; and this phenomenon probably depends upon the heat lost during the explosion in contact with so great a cooling surface, which brings the temperature of the first portions exploded below that required for the firing of the other portions. Metal is a better conductor of heat than glass; and it has been already shown that fire-damp requires a very strong heat for its inflammation.

The observation that mixtures of air and coal-gas would not explode in very narrow tubes had been previously made, unknown to Davy, by

Wollaston and Tennant. Davy likewise found that explosions would not pass through very fine wire sieves or wire gauze. He also noted that an admixture of carbonic acid and nitrogen, even in small proportions, with explosive mixtures of fire-damp greatly diminished the velocity of the inflammation.

> … It is evident then, that to prevent explosions in coal mines it is only neces-sary to use air-tight lanterns, supplied with air from tubes or canals of small diameter, or from apertures covered with wire-gauze placed below the flame, through which explosions cannot be communicated and having a chimney at the upper part, as a similar system for carrying off the foul air; and common lanterns may be easily adapted to the purpose by being made air-tight in the door and sides, by being furnished with the chimney and the system of safety apertures below and above. The principle being known, it is easy to adapt and multiply practical applications of it.

He then devised a number of lamps on this principle, and subjected them to trial with explosive mixtures in various ways. The plate below, copied from the original paper in the *Philosophical Transactions*, shows the successive forms through which the lamps passed.

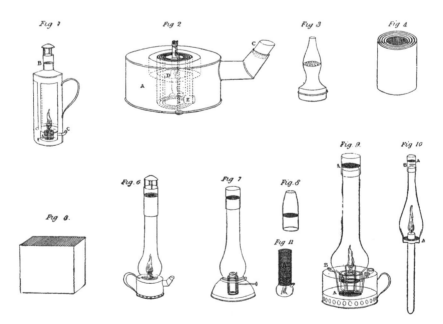

Davy's experimental safety lamps

On January 11th, 1816, he read a second paper to the Royal Society, entitled, "An account of an invention for giving light in explosive mixtures of fire-damp in coal mines by consuming the fire-damp," in which he shows that the tubes or canals as well as the sides of the lanterns may be replaced by cages or cylinders of wire gauze. The inflammable mixture will readily pass through the meshes of the gauze and will burn within it, filling the cylinder with a bright flame, but no explosion will pass outwards, even although the wire becomes heated to redness.

A fortnight later he read a third paper to the Society, "On the Combustion of Explosive Mixtures confined by Wire Gauze, with some Observations on Flame," in which he gives the results of further inquiries respecting the limits of the size of the apertures, and of the wire in the metallic gauze required to shield the flame of an oil-lamp, and describes a number of illustrations of the action of the gauze in lowering the temperature of the explosive mixture below the point of ignition. Some of these illustrations are now among the stock experiments of the lecture theatre. He offers some observations concerning the essential nature of flame, and concludes by informing the Society that his "cylinder lamps (i.e. lamps of which the flames are enclosed within a cylinder of gauze: see Fig. 11 on the facing page) have been tried in two of the most dangerous mines near Newcastle with perfect success."

The form which the lamp finally took in the hands of Mr Newman, the instrument-maker, is seen here on the right.

The trials above referred to were first made by Mr Matthias Dunn and the indefatigable Mr Hodgson in the Hebburn Colliery, and shortly afterwards by Mr John Buddle in the Wall's End Colliery. Mr Buddle has placed on record his impressions of his first experience.

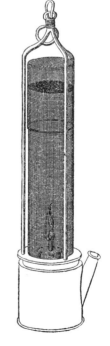

> "I first tried it," he says, "in an explosive mixture on the surface; and then took it into a mine … it is impossible for me to express my feelings at the time when I first suspended the lamp in the mine and saw it red hot … I said to those around me 'We have at last subdued this monster.'"

Some months afterwards Davy accompanied Buddle into the pit and saw his lamp in actual use.

> "Sir Humphry was delighted," says Mr Buddle, "and I was overwhelmed with feelings of gratitude to that great genius which had produced it."

Further testimony of Mr Buddle's appreciation of this memorable invention may be seen from the following extract from a letter by him to Davy. It is not only interesting in view of Davy's remark that "the evidence of the use of a practical discovery is of most value when furnished by practical men," but also as showing the rapidity with which the discovery was taken advantage of:

Walls End Colliery, Newcastle, *June 1st*, 1816.

After having introduced your safety lamp into general use in all the collieries under my direction, where inflammable air prevails, and after using them daily in every variety of explosive mixture, for upwards of three months, I feel the highest possible gratification in stating to you that they have answered to my entire satisfaction.

The safety of the lamps is so easily proved by taking them into any part of a mine charged with fire-damp, and all the explosive gradations of that dangerous element are so easily and satisfactorily ascertained by their application, as to strike the minds of the most prejudiced with the strongest conviction of their high utility; and our colliers have adopted them with the greatest eagerness.

Besides the facilities afforded by this invention to the working of coal mines abounding in fire-damp, it has enabled the directors and superintendents to ascertain, with the utmost precision and expedition, both the presence, the quantity, and the correct situation of the gas. Instead of creeping inch by inch with a candle, as is usual, along the galleries of a mine suspected to contain fire-damp, in order to ascertain its presence, we walk firmly in with the safe lamps, and with the utmost confidence prove the actual state of the mine. By observing attentively the several appearances upon the flame of the lamp, in an examination of this kind, the cause of accidents which have happened to the most experienced and cautious miners is completely developed; and this has been, in a great measure, matter of mere conjecture.

I feel peculiar satisfaction in dwelling upon a subject which is of the utmost importance, not only to the great cause, of humanity, and to the mining interest of the country, but also to the commercial and manufacturing interests of the United Kingdom; for I am convinced that by the happy invention of the safe lamp large proportions of the coal mines of the empire will be rendered available, which otherwise might have remained inaccessible, at least without an invention of similar utility, which could not have been wrought without much loss of the mineral, and risk of life and capital.

It is not necessary that I should enlarge upon the national advantages which must necessarily result from an invention calculated to prolong our supply of mineral coal, because I think them obvious to every reflecting mind; but I cannot conclude, without expressing my highest sentiments of admiration for those talents which have developed the properties, and controlled the power, of one of the most dangerous elements which human enterprise has hitherto had to encounter.

This letter is only one of many received by Davy from practical men, all telling the same story of wonder and astonishment "that so simple a look-ing instrument should defy an enemy heretofore unconquerable"; and all expressing the deepest gratitude to him as its inventor, often in language which gains in force, and even in eloquence, from its very homeliness and simple pathos.

The following address from the Whitehaven colliers was among the papers lent to me by Dr Rolleston:

September 18, 1816.

We, the undersigned, miners at the Whitehaven Collieries, belonging to the Earl of Lonsdale, return our sincere thanks to Sir Humphry Davy, for his invaluable discovery of the safe lamps, which are to us life preservers; and being the only return in our power to make, we most humbly offer this, our tribute of gratitude.

The names of eighty-two miners are appended—the majority of them—*viz.* forty-seven—with their mark (+) affixed.

What the learned world thought may be judged from the following extracts from an article in the *Edinburgh Review*—a periodical not always character-ised by a just appreciation of the work of the Royal Institution professors, for the literature of science contains few things more disingenuous or more spite-ful than the attack of "the young gentleman from Edinburgh"—afterwards known as Lord Brougham—on Thomas Young when he first made known the undulatory theory of light. In the *Review* for February, begins his article on Davy's discovery by pointing out that:

The safe lamp is a present from philosophy to the arts, and to the class of men furthest removed from the influence of science. The discovery is in no degree the effect of accident; and chance, which comes in for so large a share in the credit of human inventions, has no claims on one which is altogether the result of patient and enlightened research …

This is exactly such a case as we should choose to place before Bacon, were he to revisit the earth, in order to give him, in a small compass, an idea of the advancement which philosophy has made, since the time when he has pointed out to her the route which she ought to pursue. The great use of an immediate and constant appeal to experiment cannot be better evinced than in this example. The result is as wonderful as it is important. An invisible and impalpable barrier made effectual against a force the most violent and irresistible in its operations; and a power, that in its tremendous effects seemed to emulate the lightning and the earthquake, confined within a narrow space, and shut up in a net of the most slender texture,—are facts which must excite a degree of wonder and astonishment from which neither ignorance nor wisdom can defend the beholder. When to this we add the beneficial consequences and the saving of the lives of men and consider that the effects are to remain as long as coal continues to be dug from the bowels of the earth, it may fairly be said that there is hardly in the whole compass of art or science a single invention of which one would rather wish to be the author.

Davy was urged by several of his friends to protect his invention by a patent. Among them was Mr Buddle, who pointed out to him that he might have received his five or ten thousand a year from it.

"My good friend," was his answer, "I never thought of such a thing: my sole object was to serve the cause of humanity; and if I have succeeded, I am amply rewarded in the gratifying reflection of having done so ... More wealth could not increase either my fame or my happiness. It might undoubtedly enable me to put four horses to my carriage; but what would it avail me to have it said that Sir Humphry drives his carriage and four?"

The gratitude of some of the leading colliery proprietors for an invention so unselfishly placed at their disposal found expression in a letter from the chairman of a general meeting of the coal-owners held at Newcastle on March 18th, 1816, conveying the terms of a vote of thanks. A few months afterwards it was determined that their appreciation should take a more substantial form, and a general meeting of the coal-owners was held at Wallsend Colliery on August 31st, 1816, at which it was resolved to make Davy a present of plate.

A note of opposition was at once sounded, and it came from one of the proprietors of the Felling Colliery. Mr W. Brandling urged that it was not proved that Sir Humphry Davy was the first and true inventor of the safety lamp, or even the discoverer of the principle on which it was based.

"The conviction," he said, "upon my mind is, that Mr George Stephenson, of Killingworth Colliery, is the person who first discovered and applied the principle upon which safe lamps may be constructed; for whether the hydrogen gas is admitted through capillary tubes, or through the apertures of wire-gauze, which may be considered as merely the orifices of capillary tubes, does not, as I conceive, in the least affect the principle."

The opposition thus started very quickly gathered strength, and by appeals to local prejudice and to ignorance a degree of heat and even animosity was imported into the question, which served no other purpose than to confuse the true issue. At an adjourned meeting of the coal-owners held on October 11th, 1816, Mr William Brandling moved

> That the meeting do adjourn, until by a comparison of dates it shall be ascertained whether the merit of the safety lamp belongs to Sir Humphry Davy, or to Mr George Stephenson.

Although Mr Brandling failed to convince the meeting, it becomes necessary in the interests of truth and justice to examine the grounds upon which George Stephenson—a man of undoubted genius, and of an integrity as blameless as that of Davy, and who, as the pioneer of railway enterprise, subsequently acquired a fame as high and as deserved as that of the great chemist—has claims to be regarded as an inventor of the safety lamp. In equity, it must be admitted that the question is not merely a question of dates, for in assigning merit in a matter of this kind the calmer judgment of posterity is not wholly swayed by priority of date; it looks to circumstances, conditions, motives, and it apportions its need of approbation accordingly. The glory of Priestley as an independent discoverer of oxygen is in nowise dimmed by the circumstance that Scheele is now known to have discovered it before him. It cannot be truthfully asserted that Davy was not an independent invention of the safety lamp. What has to be determined is, has George Stephenson any such claim?

Stephenson's claim has been ably and temperately stated by Dr Smiles in his biography of George Stephenson, in *The Lives of the Engineers*, but an unbiased review of the evidence will convince most people that, however certain it may be that the Killingworth engine-tenter was an independent searcher after a method of protecting a flame, it is equally certain that he was not the discoverer of the true principle on which the safety lamp is constructed, and that the lamp associated with his name, although it bears the impress of the crude ideas with which he started, owes its real merit to the discoveries of Davy.

This controversy and the feeling it gave rise to greatly exasperated Davy, and his anger is manifested in his letters at the time. The action of the Brandlings he seemed to think was inspired by the most unworthy motives. As to his rival, he says:

> I never heard a word of George Stephenson and his lamps till six weeks after my principle of security had been published; and the general impression of the scientific men in London, which is confirmed by what I heard at Newcastle, is, that Stephenson had some loose idea floating in his mind, which he had unsuccessfully attempted to put in practice till after my labours were made known;—then, he made something like a safe lamp, except that it is not *safe*, for the apertures below are four times, and those above twenty times too large; but, even if Stephenson's plans had not been posterior to my principles, still there is no analogy between his glass exploding machine, and my metallic tissue permeable to light and air, and impermeable to flame.

On the 25th of September, 1817, as Davy passed through Newcastle on his return from Scotland, the coal-owners who had subscribed to his testimonial invited him to a banquet and presented him with the plate, which, in accordance with his wishes, took the form of a dinner-service. "I wish," he had said, "that even the plate from which I eat should awaken my remembrance of their liberality, and put me in mind of an event which marks one of the happiest periods of my life." The chairman—his friend Mr Lambton, afterwards the Earl of Durham, and who was with him under the care of Dr Beddoes at Bristol—made the presentation in an impressive and felicitous speech, and Davy acknowledged it in terms worthy of himself and of the occasion. In a subsequent speech, in response to the toast of his health, he dilated upon the theme always uppermost in his mind, and to which he never neglected the opportunity to give utterance, namely, the benefit of abstract science to mankind. He had an admirable moral to which to point, and it was driven home with all his wonted skill and power.

In what manner this plate, which was valued at about £2,500, was subsequently made subservient to the interests of science will be seen hereafter.

The friends of Stephenson were not wanting in the courage of their convictions or in determination to give substantial proof of it. In the following November they met and resolved that as in their opinion Mr G. Stephenson had been the first to discover the principle of safety and to apply it, he was entitled to some reward. Whereupon Davy's friends again assembled in public meeting on November 26th, 1817, and passed resolutions to the effect that in their opinion the merit belonged to Sir Humphry Davy alone, and that Stephenson's latest lamps were evident imitations of those of Sir Humphry Davy; and they further ordered that copies of their resolutions should appear in a number of local, London, and Edinburgh papers, and be sent to the principal owners and lessors of collieries upon the Tyne and Wear. Davy's friends in London also exerted themselves in his behalf, and a copy of resolutions similar in purport to those passed in Newcastle, signed by Sir Joseph Banks, PRS, Brande, Hatchett, and Wollaston, was sent to the newspapers.

Mr Brandling and his friends eventually collected about £800 (including 100 guineas which the meeting of October 11th had awarded Stephenson as an acknowledgment of his efforts to construct a safe lamp), and gave it, together with a silver tankard, to Mr Stephenson at a public dinner in January, 1818.

This is not the place to follow the subsequent history of the Davy lamp, or to describe the various modifications which have grown out of it, or even to show the dangers which a larger experience reveals as latent in its original form. These dangers have in great measure arisen from the development of the very system of ventilation which Buddle himself instituted; and he who in his joy exultingly exclaimed, "At last, we have subdued this monster!" has unwittingly contributed to the maleficent activity of the monster in coping with the lamp as Buddle knew it.

In the course of his numerous trials made to elucidate the principle of the safety lamp, Davy observed certain peculiarities connected with flame which led him to take up the general question. Hence arose a series of investigations, which have contributed in no small degree to our knowledge of a particularly difficult and intricate subject.

He proved, in the first place, that flame must be considered as an *explosive mixture* of inflammable gas or vapour and air, and that the heat communicated by it must depend upon its mass. The different appearance of a flame of coal gas burning in a jet in the open air, and in his safety lamp mixed with common air, led him to investigate the cause of luminosity in flame. He says:

> In reflecting on the circumstances of the two species of combustion I was led to imagine that the cause of the superiority of the light of the *stream* of coal gas might be owing to the *decomposition* of a part of the gas towards the interior of the flame, where the air was in smallest quantity, and the deposition of

solid charcoal, which, first by its *ignition,* and afterwards by its combustion, increased in a high degree the intensity of the light.

The principle of the increase of the brilliancy and density of flame by the production and ignition of solid matter explains the appearance of the different parts of the flames of burning bodies, and of the blow-pipe flame; it also explains the intensity of the light of those flames in which fixed solid matter is produced in combustion, *e.g.* phosphorus and zinc in oxygen, potassium in chlorine; and, on the other hand, the feebleness of the light of flames in which gaseous and volatile matter is alone produced, *e.g.* hydrogen in oxygen, phosphorus in chlorine. Davy's theory has not been unchallenged, but all subsequent research, when pushed sufficiently far, has shown that, as regards all ordinary illuminating flames, *i.e.* carbonaceous flames—*e.g.* coal-gas, oil, paraffin, candle—the presence of solid incandescent carbon is a prime cause of their luminosity. It had been observed that the rarefaction of a mixture of inflammable gases diminishes its combustibility: Davy proved that this diminution was not the result of the removal of pressure *per se,* but of the cooling effect thus indirectly produced. Hence, the lower the temperature of ignition of a gaseous mixture the more it may be rarefied without becoming uninflammable. In like manner he shows that by heating the gaseous mixture it may be caused to explode at a lower temperature, and that when gases combine by sudden compression, the combination is caused by the heat evolved. Also that the power of an indifferent gas to prevent the explosion of a gaseous mixture depends upon its power of abstracting heat, and that the higher the temperature of ignition of the explosive mixture the less is the amount of indifferent gas required to stop the explosion. He proved that it was quite possible to effect the gradual combination of gases without flame— that is, without the production of heat sufficient to raise the products to incandescence; and he discovered the singular fact that platinum would induce the combination of many inflammable gases and vapours, and on this circumstance based the construction of his flameless lamp.

In the early summer of 1818, he thus wrote to his mother:

> My dear Mother,—We are just going on a very interesting journey. I am first to visit the coal miners of Flanders, who have sent me a very kind letter of invitation and of thanks for saving their lives. We are then going to Austria,

where I shall show Vienna to Lady Davy, and then visit the mines; and lastly, before I return, we are going to visit Naples.

I have the commands of his Royal Highness the Prince Regent to make experiments upon some very interesting ancient manuscripts, which I hope to unfold. I had yesterday the honour of an audience from his Royal Highness, and he commissioned me to pursue this object in the most gracious and kind manner …

We shall be absent some months. With kindest love to my sisters and my aunts,

I am, my dear mother,

Your most affectionate son,
H. DAVY.

A few months after this visit to the Prince Regent he received the intimation that he had been created a baronet.

He arrived at Naples in the autumn, and began his researches on the Herculaneum manuscripts referred to in his letter. His first results were sufficiently encouraging to induce him to make some prolonged experiments with a view of discovering a method of unfolding them. He found that the papyri had suffered not so much from fire, as was believed, as from a gradual change in vegetable structure, similar to that which accompanies the transformation of vegetable matter into lignite. He managed to unroll a number, and an account of his results was communicated to the Royal Society in 1821. But from the fragmentary character of the papyri these were found to be of little value to literature. Subsequently difficulties were put in his way by the curators, of the museum, and ultimately his investigations were abandoned, not without some little exhibition of temper and resentment on his part.

During his stay at Naples he again interested himself in the volcanic phenomena of Vesuvius, and his observations constitute the material of a paper which was published in the *Philosophical Transactions* in 1827, and many of his personal experiences in connection with the subject are referred to in his last work, *Consolations in Travel*.

He left Naples in the spring of 1819, and after a short stay at the baths of Lucca he went for the summer and early autumn into the Tyrol, whence he again proceeded to Lucca, and on the approach of winter returned to Naples, where he arrived on December 1st. He quitted it in the spring of 1820, and travelled slowly home by the south of France and Bordeaux, arriving in England about the middle of June. On the 19th of that month Sir Joseph Banks died, and so terminated his forty-two years' presidency of the Royal Society, to which position he was elected before Davy was even born. Davy immediately announced his intention of becoming a candidate for the vacant chair, and was elected at the following anniversary meeting on November 30th.

XI

DAVY AND THE ROYAL SOCIETY—HIS LAST DAYS

D AVY WAS ELECTED INTO THE Royal Society in 1803. His certificate describes him as "a gentleman of very considerable scientific knowledge, and author of a paper in the *Philosophical Transactions*." Two years afterwards—that is, in his twenty-seventh year—he was awarded the Copley medal; from which we may infer either that the Society considered their medal not to have the lustre it now possesses, or that they had a confident belief in the power and coming greatness of the recipient, since the papers for which it was given are perhaps the least meritorious of Davy's productions. His active interest in the affairs of the Society led to his election—or rather selection, for the appointment in those days was made by the President—as one of the Secretaries, a position he held until 1812, when he resigned it at the time of his marriage. In 1816 he received the Rumford medal of the Society for his work in connection with flame and the safety-lamp—an award which would have given a peculiar satisfaction to Rumford had he lived to witness it.

On the death of Sir Joseph Banks the general voice of the Fellows seemed to designate Wollaston as his successor. It was, indeed, Sir Joseph Banks's desire that Dr Wollaston should be nominated. "So excellent a man," he remarked to Barrow, "of such superior talents, and everyway fitted for the situation. Davy is a lively and talented man, and a thorough chemist; but ... he is rather too lively to fill the chair of the Royal Society with that degree of gravity which it is most becoming to assume." Oh this gravity! "*La gravité*," says La Rochefoucauld, "*est un mystère du corps, inventé pour cacher les défauts de l'esprit.*" And Sir Joseph had enough of it and to spare. Wollaston—a man of wide knowledge, steady, cautious, and sure,—of cool judgment and sagacious views, as Davy said of him—felt no inclination to accept a position for which his retiring habits and reticent disposition to some extent unfitted him, and he declined to be put in nomination. Davy's attitude is indicated in the following letter to his friend Poole:

> I feel that the President's chair, after Sir Joseph, will be no light matter; and unless there is a strong feeling in the majority of the body that I am the most proper person, I shall not sacrifice my tranquillity for what cannot add to my reputation, though it may increase my power of being useful.
>
> I feel it a duty that I owe to the Society to offer myself; but if they do not feel that they want me, (and the most active members, I believe, do) I shall not force myself upon them.

The "strong feeling in the majority" was shown on the day of election. A few votes were given in favour of Lord Colchester, but Davy's triumph was practically complete.

He thus writes to Mr Poole in answer to a letter of congratulation:

> I have never needed any motive to attach me to science, which I have pursued with equal ardour under all circumstances, for its own sake, and for the sake of the pubic, uninfluenced by the fears of my friends, or the calumnies of my enemies. I glory in being in the chair of the Royal Society, because I think it ought to be a reward of scientific labours, and not an appendage to rank or fortune; and because it will enable me to be useful in a higher degree in promoting the cause of science.

Davy was re-elected to the Presidential Chair without opposition for seven successive years—until, in short, his failing health compelled him to resign. Although the Society owes much to him, he himself derived little satisfaction or pleasure from the position. He soon found, as he anticipated, that the President's Chair, after Sir Joseph, was no light matter; and there is little doubt that the worries and cares of the office contributed to his untimely death. In bearing, manner, temperament—in fact, in almost every particularity —he was the very opposite to his predecessor; and when the discontent which had slumbered, with an occasional awakening, during Sir Joseph's long reign, and which his firmness, tact, and the weight of his personal character had for the time allayed, broke out, Davy was too impulsive and irascible to deal with it as Banks had done, and matters which a less sensitive or a more impassive man would have unheeded were causes of annoyance and ill-temper to him, and served to add to the spirit of disunion which prevailed. But if he occasionally lacked discretion, he was never wanting in zeal. He laboured incessantly to add to the dignity and usefulness of the Society. He strove in every way to enhance the character of its publications and to raise the standard of Fellowship. His great ambition was to bring the Society into more intimate relation with the State.

"It was his wish," says his brother, "to have seen the Royal Society an efficient establishment for all the great practical purposes of science, similar to the college contemplated by Lord Bacon, and sketched in his New Atlantis; having subordinate to it the Royal Observatory at Greenwich for astronomy; the British Museum, for natural history, in its most extensive acceptation."

Realising in his own case what such a laboratory as that of the Royal Institution, supported wholly by private liberality, had done for science, it was his desire that similar laboratories, amply provided with all means requisite for original inquiry, should be maintained and administered by the Society. But, as his brother adds, the Government, although ready enough to consult him when in want of his knowledge or of that of other Fellows of the Society, was lukewarm and indifferent in of science, and he received no effectual support. It is true that towards the end of his Presidency the Society received a mark of Royal favour by the foundation of the Royal Medals in 1825, but from various causes the medals were not actually forthcoming until 1833, when the Duke of Sussex was in the Chair, although no fewer than ten awards had been made in the meantime. In his attention to the personal duties of his office Davy was unremitting; His addresses were a feature of the session; in these he displayed all the ardour, eloquence and poetical fervour, and, it may be added, all the egoism, which characterised his lectures. He delighted to dwell upon the power and dignity of science, its worth as a mental instrument, and its value to the national life. In his announcements of the awards of the Society's medals the range of his knowledge, his power of exposition, and his faculty of felicitous expression found ample opportunity for exercise. He was the first President to introduce obituary notices of Fellows, and his *éloges* are marked by judgment, taste, and warmth of feeling.

In everything that related to the dignity and ceremony of his office he was, as might have been expected, most punctilious. Although as a rule somewhat careless in dress, he invariably took the chair in full Court dress, sitting covered, and with the mace of office—the veritable "bauble" which Cromwell ordered to be removed from the table of the House of the Commons—in front of him, as is still the custom.

To enhance his dignity we are told that he petitioned Government for the Red Ribbon of his predecessor, and it was said that he felt so certain his request would be granted that his name was printed with the coveted letters K.B. appended.

During the session he followed the practice of Sir Joseph Banks in assembling the Fellows at a weekly conversazione at his house in Lower Grosvenor Street. Subsequently, on his removal to Park Street, these meetings were held in the apartments of the Society at Somerset House. Davy's vivacity and conversational powers made the gatherings in the outset a great success, but when the tide of his unpopularity as President set in, the attendance fell off, and they were eventually discontinued.

During the autumn preceding his first election he spent some time with Scott at Abbots ford, in company with Wollaston and Mackenzie (the Man of Feeling), and Lockhart gives some account of him as the party started on a sporting expedition on a September morning.

> But the most picturesque figure was the illustrious inventor of the safety lamp. He had come for his favourite sport of angling … and his fisherman's costume—a brown hat with flexible brims, surrounded with line upon line, and innumerable fly-hooks; jack-boots worthy of a Dutch smuggler, and a fustian surtout dabbled with the blood of salmon—made a fine contrast to the smart jackets, white-cord breeches, and well polished jockey-boots of the less distinguished cavaliers about him. Dr Wollaston was in black, and with his noble serene dignity of countenance might have passed for a sporting archbishop … I have seen Sir Humphry in many places, and in company of many different descriptions; but never to such advantage as at Abbotsford. His host and he delighted in each other, and the modesty of their mutual admiration was a memorable spectacle. Davy was by nature a poet—and Scott, though anything but a philosopher in the modern sense of that term, might, I think it very likely, have pursued the study of physical science with zeal and success, had he happened to fall in with such an instructor as Sir Humphry would have been to him, in his early life. Each strove to make the other talk—and they did so in turn more charmingly than I have ever heard either on any other occasion whatsoever. Scott in his romantic narratives touched a deeper cord of feeling than usual, when he had such a listener as Davy and Davy, when induced to open his views upon any question of scientific interest in Scott's presence, did so with a degree of clear energetic eloquence, and with a flow of imagery and illustration, of which neither his habitual tone of table-talk (least of all in London), nor any of his prose writings (except, indeed, the post-

humous Consolations in Travel) could suggest an adequate notion. I say his prose writings—for who that has read his sublime quatrains on the doctrine of Spinoza can doubt that he might have united, if he had pleased, in some great didactic poem, the vigorous ratiocination of Dryden and the moral majesty of Wordsworth? I remember William Laidlaw whispering to me, one night, when their 'wrapt talk' had kept the circle round the fire until long after the usual bed-time of Abbotsford—'Gude preserve us! This is a very superior occasion! Eh, sirs!' he added, cocking his eye like a bird, 'I wonder if Shakspeare and Bacon ever met to screw ilk other up?

In spite of the many calls upon his time and energies entailed by his duties as President, he still found opportunity to work in his laboratory, and one outcome of his labours was a paper "On the magnetic phenomena produced by electricity," published in the *Philosophical Transactions* for 1821—the sequel of a letter addressed to Wollaston and also printed in the Transactions. This memoir was followed a few months later by a communication "On the Electrical phenomena exhibited *in vacuo.*"

These papers, together with one on a New Phenomenon of Electro-Magnetism, published in 1823, are interesting in relation to the development of Oersted's great discovery, and in connection with the subsequent work of Faraday.

With that power of generalisation which is one of the distinguishing marks of his genius, he shows the possible connection of the facts he had observed with the phenomena of terrestrial magnetism. He concludes his first paper by asking

> whether the magnetism of the earth may not be owing to its electricity, and the variation of the needle to the alterations in the electrical currents of the earth, in consequence of its motions, internal changes, or its relations to solar heat; and whether the luminous effects of the auroras at the poles are not shown, by these new facts, to depend on electricity. This is evident, that if strong electrical currents be supposed to follow the apparent course of the sun, the magnetism of the earth ought to be such as it is found to be.

It is perhaps idle to speculate on such a matter, but it is more than likely that had Davy been free from the cares and restraints of office, and from the innumerable distractions inseparable from his position in the social and scientific world of London, he might have revealed the possibilities in electro-magnetism with the same brilliant success as he had done those of voltaic electricity. He was now at the maturity of his mental power, and had still much of the enthusiasm and ardour which characterised his earliest work, and under serener conditions he might have achieved triumphs not less striking than those reserved for Faraday. His few short papers on the subject indicate that he fully realised the great wealth of the new territory thus opened out to science; and into which he was one of the first to penetrate. But it is sad to think that he might have extended a more generous hand to one who, equally with himself, was striving to enter the new land, and who eventually did enter and for a time possessed it. In the concluding words of Davy's last paper on electro-magnetism, we discern in the allusion to Wollaston's idea of the possibility of the rotation of the electro-magnetic wire round its axis "the rift within the lute" in his relations towards his assistant, which widened in the matter of the condensation of chlorine, and which threatened to become an open breach when Faraday was elected into the Royal Society.

The jealousy thus manifested by Davy is one of the most pitiful facts in his history. It was a sign of that moral weakness which was at the bottom of much of his unpopularity, and which revealed itself in various ways as his physical strength decayed.

Greedy as he was of fame—that infirmity of noble minds—many incidents in his life up to this period prove that he was not wanting on occasion in a generous appreciation of the work of his contemporaries, even in fields he might reasonably claim as his own. But, although in his intellectual combats he could show at times a certain knightly courtesy, it must be confessed that he was lacking in the magnanimity which springs from the charity that envieth not.

In genius he was unquestionably superior to Faraday; in true nobility of character he was far below him. It is almost impossible to avoid comparing him with Faraday. Indeed it is one of the penalties of his position that he has to be tried by so severe a standard, and it may well be that his good name, which, as Bacon says, is the proper inheritance of the deceased, has suffered unduly in consequence. His true place in the history of science is defined by his discoveries; it is a sad reflection that the lustre of his fame has been dimmed rather than heightened by what has been styled the greatest of them all—Faraday. But there has undoubtedly been injustice in the comparisons which have been made. What Davy was to Faraday, Faraday would have been the first to admit. Davy made himself what he was by the sheer force of his unaided genius; what Faraday became was in large measure due to his connection with Davy, and the germs of his greatest works are to be traced to this association. This fact has been frankly acknowledged by Faraday. To the end of his days he regarded Davy as his true master, preserving to the last, in spite of his knowledge of the moral frailties of Davy's nature, the respect and even reverence which is to be seen in his early lecture notes and in his letters to his friend Abbott. Faraday was not easily roused to anger, but nothing so effectually moved him as any aspersion of Davy's character as a man of science, or any insinuation of ungenerous treatment of himself by Davy.

At about this time—that is, in the autumn of 1823—Davy gave the first signs of the obscure malady which ultimately occasioned his death. In a letter to his brother, in which he describes his symptoms, we have a reference, also, to his domestic worries: "To add to my annoyances, I find my house, as usual, after the arrangements made by the mistress of it, without female servants; but in this world we have to suffer and bear, and from Socrates down to humble mortals, domestic discomfort seems a sort of philosophical fate."

He was able, however, to continue his scientific work, but instead of the fame and applause on which he so confidently counted, he found only disappointment and chagrin.

In 1823 the Admiralty sought the advice of the Royal Society as to "the best means of securing to the service copper of the most durable quality, and such as will preserve the smoothest surface." A committee of the Society was appointed under Davy's direction, to consider the question, which ultimately resolved itself into one of preventing the corrosion of the metal. In this matter Davy's special experience proved most useful, and, as a fact, he took all the experimental part of the inquiry upon himself, and with what result may be seen from the following letter to his brother:

> Firle, *Jany.* 30, 1824.
>
> I have lately made a discovery of which you will for many reasons be glad. I have found a complete method of preserving the copper sheeting of ships, which now readily corrodes. It is by rendering it negatively electrical. My results are of the most beautiful and unequivocal kind: a mass of tin renders a surface of copper 200 or 300 times its own size sufficiently electrical to have no action on sea water.
>
> I was led to this discovery by principle, as you will easily imagine; and the saving to government and the country by it will be immense. I am going to apply it immediately to the navy. I might have made an immense fortune by a patent for this discovery, but I have given it to my country; for in everything connected with interest, I am resolved to live and die at least 'sans tâche.'

His method of rendering the copper negatively electric consisted in affixing to the sheets a number of short bars of iron or zinc, properly curved to the shape of the vessel. In this way the "protectors," as the zinc or iron bars were called, gradually corroded, whilst the copper remained unattacked. But, as Dr Paris remarks, the truth of the theory was completely established by the failure of the remedy. The ship's bottom became so foul by the adhesion of shells and weed that her speed was greatly impeded, and after a number of trials, in the course of which a steam vessel was placed at his disposal, in which he made a voyage to Norway and back, the Admiralty directed the protectors to be removed. To add to his mortification, the order was issued immediately after a communication to the Royal Society announcing the complete success of his plan. Throughout the whole of this business he was exposed to a number of vexatious attacks, which greatly embittered him and reacted disastrously upon his health and character. So long as there was the hope of success and the prospect of reward his claims to the originality of the invention were contested; no sooner was the project abandoned than he was assailed in the periodical press and made an object of sarcasm and censure. As might be imagined, his philosophy was not proof against such attacks. He wrote to his friend Children:

> A mind of much sensibility might be disgusted, and one might be induced to say why should I labour for public objects, merely to meet abuse I—I am irritated by them more than I ought to be; but I am getting wiser every day— recollecting Galileo, and the times when philosophers and public benefactors were burnt for their services.

During the autumn his indisposition increased, and his home letters show that the wonderful elasticity of spirit, which, as his brother remarks, had hitherto carried him lightly and joyously through life, over all its rubs and cares, now seemed to flag. He had an ailing winter, and with the spring came news of his mother's illness. He could only write with difficulty:—"If it please God, I will certainly be at Penzance the last week in October or the first in November." He never saw her again; she rallied for a time, but died somewhat suddenly in September. Davy never really recovered from the shock of her death. It was with the greatest difficulty that he was able to preside at the anniversary meeting of the Society on the ensuing St Andrew's Day. The effort was so marked that those near him feared he was on the verge of apoplexy, and he was too ill to attend the dinner. A few weeks later he had a slight attack of paralysis, from which he only slowly recovered. His good friend Dr Babington[1] ordered him abroad, away from "the convivial epicurean habits of London society," and from "the many annoyances and causes of injurious excitement to which he was exposed at home." He set out with his brother John, in the depth of winter—"a dreary beginning of a dreary journey." He avoided Paris; he would not even pass through it, so apprehensive was he that he should not escape from "the allurement—or, rather, excitement—of its society" if he stopped there. The roads were in a wretched state, the country covered with snow, and "no object to arrest the eye, except a village here and there rising out of the white waste, or a distant steeple, or some solitary tree." The cold was intense, and once or twice the travellers were benighted, the wheels of their carriage being locked in the frozen ruts. As they passed through the towns Davy, who seemed to cling to life with a passionate tenacity, would visit the churches, and, falling on his knees, would offer up a silent prayer. They crossed Mont Cenis in a storm of wind and amidst drifting snow, and with great difficulty got down to Susa on sledges. The snow in Lombardy was deeper than in the passes of the Alps, and even at Ravenna, where they arrived in the first week of March, it was still to be seen in the ditches. Here his brother left him, his duties as an army surgeon calling him to Corfu. In spite of severe weather, the discomforts of travelling at such a time, and the forced delays at wretched inns, Davy gradually improved; his brother noted before he left that he was certainly stronger, less paralytic, and more active. He wrote to his friend Poole:

> I am, thank God, better, but still very weak, and wholly unfit for any kind of business and study. I have, however, considerably recovered the use of all the limbs that were affected; and as my amendment has been slow and gradual, I hope in time it may be complete. But I am leading the life of an anchorite, obliged to abstain from flesh, wine, business, study, experiments, and all things that I love; but this discipline is salutary, and for the sake of being able to do something more for science, and I hope for humanity, I submit to it, believing that the Great Source of intellectual being so wills it for good.

He tells Poole that he had chosen Ravenna—this spot of the declining Empire of Rome—as one of solitude and repose, and as out of the way of travellers and in a good climate. He was interested, too, in its many associations with his friend Byron, with Dante, and in its old-world memories of Theodoric and his lost legions. How the place affected him in his state of physical enfeeblement, but with his mind chastened and purified, may be seen in the character of much that he wrote there, and particularly in his poems, with their many notes of sadness and hope, trust and resignation. He was lodged in the Apostolical Palace by the kindness of the Vice-Legate—a graceful, learned, and accomplished man, with whom he contracted a warm friendship. He says he could not speak of his goodness without tears of gratitude. in his eyes, and with this exception and an occasional visit from the Countess Guiccioli he had no society. Most of his time was spent in riding amidst the pines and junipers, or following the petzardone among the marshes of La Classe; or in reading and in the study of natural history.

> The natural strength of his mind," says his brother, "was very clearly manifested under these circumstances. Dependent entirely on his own resources; no friend to converse with; no one with him to rely on for aid, and in a foreign country, without even a medical adviser; destitute of all the amusements of society; without any of the comforts of home—month after month, he kept on his course, wandering from river to river, from one mountain lake and valley to another, in search of favourable climate; amusing himself with his gun and rod, when sufficiently strong to use them, with '*speranza*' for his rallying word.

With the approach of spring he passed by way of Gorizia into Illyria, and, as the heat increased, into Upper Austria, Bavaria, and Switzerland, and back, in the late summer, to Illyria. His journals give a fairly full account of his movements and of the manner in which he spent his time; they also indicate his state of mind, the alternations of hope and despondency, and his constant struggles with the insidious disease which was gradually exhausting his physical powers.

He wrote to his wife from Laybach:

> You once *talked* of passing *this* winter in Italy; but I hope your plans will be entirely guided by the state of your health and feelings. Your society would undoubtedly be a very great resource to me, but I am so well aware of my own present unfitness for society, that I would not have you risk the chance of an uncomfortable moment on my account. I often read Lord Byron's Euthanasia: it is the only case, probably, where my feelings perfectly coincide with what his were.

At times the feeling of despair was so intense that he actually seemed apprehensive of suicide. It was probably under the influence of such a fear that he wrote in his journal that he had too strong a faith in the optimism of the system of the universe ever to accelerate his dissolution.

> I have been and am taking a care of my health which I fear it is not worth; but which, hoping it may please Providence to preserve me for wise purposes, I think it my *duty.*

On another occasion he wrote to Lady Davy:

I am glad to hear of your perfect re-establishment, and with health and the society of London, which you are so well fitted to ornament and enjoy, your '*viva la felicità*' is much more secure than any hope belonging to me.

Subsequently he wrote:

Should your feelings or inclination lead you *to the land of the sun*, I need not say what real pleasure it would give me to enjoy your society; but do not make any sacrifice on my account.

A couple of days afterwards he wrote:

I hope I shall have the delight of seeing you at Baden Baden. If not, I shall come to England … Pray let my physicians know what an obedient patient I am … God bless you, my dear Jane!

Towards the end of September, and at Baden, the solitary man wrote:

I fear my light of life is burnt out, and that there remains nothing but stink, and smoke and dying snuff … *Dubito fortissime restaurationem meum.*—Decidedly worse and have decided to go home immediately.

At Mayence he informed his wife that he trusted soon to see her in Park Street. He had a lingering hope that she might still be induced to cross the water, and that he might meet her at Calais.

I think you will find me altered in many things—with a heart still alive to value and reply to kindness, and a disposition to recur to the brighter moments of my existence of fifteen years ago, and with a feeling that though a burnt-out flame can never be rekindled, a smothered one may be … I hope it is a good omen that my paper by accident is *couleur de rose*.

He had previously determined to resign the chair of the Royal Society, and announced his decision in a letter to his old friend Davies Gilbert, the treasurer. To his wife he wrote:

If I had perfectly recovered I know not what I should have done with respect to the P. under the auspices of a new and more enlightened government; but my state of health renders the resignation *absolutely* necessary. To attempt business this year would be to prepare for another attack.

He is pleased with the idea that Sir Robert Peel, who had "no scientific glory to awaken jealousy," may be his successor; and he resumes:

> The prosperity of the Royal Society will always be very dear to me, and there is no period of my life to which I look back with more real satisfaction than the six years of labour for the interests of that body. I never *was*, and never could be, unpopular with the active and leading members, as six unanimous elections proved; but because I did not choose the Society to be a tool of Mr ——'s journal jobs, and resisted the admission of improper members, I had some enemies, who were listened to and encouraged from Lady ——'s chair. I shall not name them, but as Lord Byron has said 'my curse shall be forgiveness.'

He arrived in London in the first week in October, and towards the end of the month he wrote to his friend Poole that he had consulted all the celebrated men who had written upon or studied the nervous system.

> They all have a good opinion of my case, and they all order absolute repose for at least twelve months longer, and will not allow me to resume my scientific duties or labours at present; and they insist upon my leaving London for the next three or four months and advise a residence in the West of England.

Poole promptly asked him down to Nether Stowey. His friend relates that although his bodily infirmity was very great and his sensibility painfully acute—("Here I am, the ruin of what I was!" he exclaimed on his arrival)—his mind still showed much of its wonted ardour and vigour. He spent his mornings in literary work, mainly on his *Salmonia; or, Days of Fly-fishing*, a philosophical disquisition on angling, published in 1828, and which, despite the rollicking banter of Christopher North, passed through five or six editions. Davy had the ambition to do for fly-fishing what Walton had done for the humbler art of bottom-fishing. But Davy's book, although constructed on much the same lines as *The Compleat Angler*, lacks every feature which has made honest Izaak's work immortal—the quaint simplicity, the homely wit, the delicate humour, the delightful charm—the reflection, in a word, of the mental features of a lovable man blessed with the ornament of a meek and quiet spirit. The egotism and garrulity of Piscator are delicious; the loquacity and self-confidence of Davy's Halieus are tiresome to the last degree. We are bored with his long didactic speeches, his consciousness of superiority, and his cheap and tawdry senti-ment. It was a poor return for all the kindness and skill of Babington, that his patient should have seen in such a creation the character of one of the most charming and estimable of men.

More than one mention has been made in this biography of what Maria Edgeworth termed Davy's "little madness." Indeed, the love of angling amounted to a passion with him; and he told Ticknor that he thought if he were obliged to renounce either fishing or philosophy he should find the struggle of his choice pretty severe. Whenever he could escape from town he would hie him to some favourite stream and spend the day in the practice of his beloved art. He was known to have posted a couple of hundred miles for the sake of a day's fishing, and to have returned contented, although he had never a rise. When confined to Albemarle Street, and chafing at his inability to get away, he would sometimes turn over the leaves of his fly-book and derive much consolation from the sight of his hackles and harles, his green-tails, dun cuts, red spinners, and all the rest of the deadly paraphernalia associated in his mind with the memories of pleasant days and exciting combats. He greatly prided himself on his skill, and his friends were often secretly amused to notice his ill-concealed chagrin when a brother-angler outvied him in the day's catch or in the narration of some piscatorial tri-umph. They were amused, too, at the costume which he was wont to don on such occasions—his broad-brimmed, low-crowned hat, lined with green and garnished with flies; his grey-green jacket, with a multitude of pockets for the various articles of his angling gear; his wading-boots and knee-caps—all made up an attire as original as it was picturesque. In these fishing expeditions he enjoyed some of the happiest hours of his life; at such times he threw off his cares and annoyances; he was cheerful even to hilarity, and never was his conversation more sprightly or more entertaining.

In spite of the thoughtful care of his friend Poole, Davy's health showed no material improvement, and at times his feeling of despondency was very great. His confidence in his mental powers, however, never forsook him. He said on one occasion:

> I do not wish to live, as far as I am personally concerned; but I have views which I could develope, if it please God to save my life, which would be useful to science and to mankind.

"His inherent love of the laboratory (if I may so speak)," says Mr Poole, "was manifested in a manner which much interested me at the moment. On his visiting with me a gentleman in this neighbourhood who had offered to let him his house, and who has an extensive philosophical apparatus, particularly complete in electricity and chemistry, he was fatigued by the journey; and as we were walking round the house very languidly, a door opened, and we were in the laboratory. He threw his eyes round the room, which brightened in the action—a glow came over his countenance, and he appeared himself twenty years ago. He was surprised and delighted and seemed to say, 'This is the beloved theatre of my glory.' I said 'You are pleased.' He shook his head and smiled."

In the spring he determined to quit England for his beloved Illyria, and towards the end of May arrived by easy stages at Wurzen. In his journal he wrote:

May 22. To my old haunt, Wurzen, which is sublime in the majesty of Alpine grandeur; the snowy peaks of the Noric Alps rising above thunder clouds, whilst spring in all its bloom and beauty blooms below; its buds and blossoms adorning the face of Nature under a frowning canopy of dark clouds, like some Judith beauty of Italy—a Transteverene brow and eye, and a mouth of Venus and the Graces.

From Aussee he wrote to his brother:

It suits me better to wile away my days in this solitary state of existence, in the contemplation of Nature, than to attempt to enter into London society, where recollections call up the idea of what I was, and the want of bodily power teaches me what a shadow I am ... I am now going to Ischl, where there are warm salt baths to try if they will renovate the muscular powers of my arm and leg ... I wish to go to Trieste in October, to make the experiments I have long projected on the torpedo.

He derived some little benefit from the treatment at Ischl, and in October went to Trieste, where he carried out his projected experiments on the electricity of the torpedo, the results of which he communicated to the Royal Society. This paper was the last of his scientific memoirs. In the middle of November he arrived at Rome, where he learnt that Wollaston also had been stricken with paralysis.[2] On February 6th, 1829, he wrote to Poole:

I am here *wearing away* the winter,—a ruin amongst ruins! … I hope you got
a copy of my little trifle 'Salmonia.' … I write and philosophise a good deal,
and have nearly finished a work with a higher aim than the little book I speak
of above, which I shall dedicate to you. It contains the essence of my philosophical opinions, and some of my poetical reveries. It is like the 'Salmonia,' an
amusement of my sickness; but *'paulo majora canamus.'* I sometimes think of
the lines of Waller, and seem to feel their truth—

> 'The soul's dark cottage, batter'd and decay'd,
> Lets in new light through chinks that Time has made.'

The work to which he here alludes, and which he did not live to see printed,
was his *Consolations in Travel; or, The Last Days of a Philosopher.* He had practically finished it at the date of his letter, and had written in his journal: "*Si
moro, spero che ho fatto il mio dovere, e che mia vita, non e stato vano ed inutile.*"
On February 20th he was seized with a new attack, and his right side was
quite powerless. On the 23rd he dictated the following letter to his brother,
who was then at Malta:

Notwithstanding all my care and discipline, and ascetic living, I am dying from
a severe attack of palsy, which has seized the whole of the body with the exception of the intellectual organ … the weakness increases and a few hours or days
will finish my mortal existence. I shall leave my bones in the Eternal City. I
bless God that I have been able to finish all my philosophical labours … I hope
you will have the goodness to see these works published … I have given you,
by a codicil to my will, the copyright of these books … God bless you, my dear
John! May you be happy and prosperous!

The letter was signed by him, and he added in his own handwriting, only
just legible, "Come as quickly as possible."

Two days afterwards he dictated another letter, in which he gives minute
directions concerning some experiments on the torpedo which he wished his
brother to make. He describes the apparatus which may be employed and
indicates where the torpedoes may be procured; and he concludes: "Pray do
not neglect this subject, which I leave to you as another legacy." It was the
16th of March before Dr Davy could reach Rome. The stricken man's pale
and emaciated countenance lighted up as he saw his brother at his bedside.
He spoke as if he had only a few hours to live, and rejected all expectation and
hope of recovery, saying he was sure his career was run.

Under the care and medical skill of Dr Davy, however, he rallied.

"As he mended," says his brother, "the sentiment of gratitude to Divine Providence was overflowing, and he was most amiable and affectionate in manner. He often inculcated the propriety, in regard to happiness, of the subjugation of self, in all selfishness, as the very bane of comfort, and the most active cause of the dereliction of social duties, and the destruction of good and friendly feelings; and he expressed frequently the intention, if his life were spared, of devoting it to purposes of utility (seeming to think lightly of what he had already done), and to the service of his friends, rather than to the pursuits of ambition; pleasure, or happiness, with himself for their main object.

But, Dr Davy adds:

Now that he was intent on recovery, he no longer took the same interest in *my* examination of the torpedo, as if he looked forward to the time when *he* should be able to enter into the investigation actively again.

At the beginning of April Lady Davy arrived from England, and he had so far improved that it was decided to remove him to Geneva. By easy stages, and occasional halts of two or three days at the more interesting places, he arrived at Geneva on May 28th. He bore the journey well: the delightful freshness of the spring, the bursting vegetation, the many streams, the pure mountain air, and the indescribable influence of Alpine scenery, seemed to invigorate him. On his arrival at the inn ("La Couronne") he walked to the window, looked out upon the lake, and expressed a longing wish to throw a fly upon its blue waters. Lady Davy here broke to him the news of the death of his old friend and colleague, Thomas Young. This, coming so soon after the loss of Wollaston, profoundly affected him. During the evening he struck his elbow against the projecting arm of the sofa on which he sat; the blow gave him great pain, and seemed to have the most extraordinary effect. He was got to bed as soon as possible. He took an anodyne, and desired to be left alone. Soon after midnight he was found to be insensible, and shortly before three on the morning of the 29th of May he died. In his will he had enjoined that he should be buried where he died: *Natura curat suas reliquas*, he had written.

The City gave him a public funeral, and representatives of every institution in the town followed his remains to their resting-place in the cemetery at Plain Palais. A simple monument, with the following inscription, marks the spot:

Hic jacet
HUMPHRY DAVY
Eques Magnæ Britanniæ Baronetus
Olim Regiæ Societ. Londin. Præses
Summus Arcanorum Naturæ Indigator.
Natus Penzantiæ Cornubiensum XVII Decemb. MDCCLXXVIII.
Obit Genevæ Helvettorum XXIX MAI MDCCCXXIX.

His widow placed a tablet to his memory in the north transept of Westminster Abbey. His baronetcy died with him. By his will he directed that the service of plate given to him by the coal-owners should, after Lady Davy's death, pass to his brother, and that in the event of his having no heirs in a position to make use of it, it should be melted and given to the Royal Society, "to found a medal to be given annually for the most important discovery in chemistry anywhere made in Europe or Anglo-America." This is the origin of the Davy Medal which has been awarded annually by the Society since 1877.

Many eloquent tributes have been paid to the genius and labours of Davy, and some of these eulogies are among the most brilliant passages in the literature of science. One of the best-known is from the gifted pen of Dr Henry in the preface to his "Elements of Chemistry," published soon after Davy's death. He thus sketches the more striking characteristics of the great chemist.

> "Davy," he says, "was imbued with the spirit, and was a master of the practice, of the inductive logic; and he has left us some of the noblest examples of the efficacy of that great instrument of human reason in the discovery of truth. He applied it not only to connect classes of facts of more limited extent and importance but to develope great and comprehensive laws, which embrace phenomena that are almost universal to the natural world. In explaining these laws, he cast upon them the illuminations of his own clear and vivid conceptions;—he felt an intense admiration of the beauty, order and harmony which are conspicuous in the perfect chemistry of Nature;—and he expressed these feelings with a force of eloquence which could issue only from a mind of the highest powers and of the finest sensibilities."

Not less forcible or eloquent, although hardly so well known, is the estimate in Silliman's *American Journal of Science and Arts* for January, 1830. After an analysis of Davy's mental attributes the writer concludes:

We look upon Sir Humphry Davy as having afforded a striking example of what the Romans called *a man of good fortune*; —whose success, even in their view, was not however the result of accident, but of ingenuity and wisdom to devise plans, and of skill and industry to bring them to a successful issue. He as fortunate in his theories, fortunate in his discoveries, and fortunate in living in an age sufficiently enlightened to appreciate his merits;—unlike, in this last particular to Newton, who (says Voltaire), although he lived forty years after the publication of the *Principia*, had not, at the time of his death, twenty readers out of Britain. Some might even entertain the apprehension that so extensive a popularity among his contemporaries is the presage of a short-lived fame; but his reputation is too intimately associated with the eternal laws of Nature to suffer decay; and the name of Davy, like those of Archimedes, Galileo and Newton, which grow greener by time, will descend to the latest posterity.

Such, then, is the story of a life of fruitful endeavour and splendid achievement;—the record of one who, if not wholly good or truly noble, has, left a track of greatness in his passage through the world.

1. "Babington, the best and warmest-hearted friend, the kindest husband and father, and perhaps the most disinterested physician of his time; with good talents, and a fine tact, and a benevolence which created sympathy for him wherever he appeared, and I believe often cured his patients."

2. He died on December 22nd, 1828.

Also available from Nonsuch Publishing

For forthcoming titles and sales information see
www.nonsuch-publishing.com